P9-CKF-906

101 Things You Don't Know About Science and No One Else Does Either

101 Things You Don't Know About Science and No One Else Does Either

James Trefil

A Mariner Book
HOUGHTON MIFFLIN COMPANY
BOSTON • NEW YORK

For information about permission to reproduce selections from this book, write to
Permissions, Houghton Mifflin Company,
215 Park Avenue South, New York,
New York 10003.

Library of Congress Cataloging-in-Publication Data
Trefil, James S., date.
[Edge of the unknown]
101 things you don't know about science and
no one else does either / James Trefil.
   p.   cm.
Originally published: The edge of the unknown.
Boston : Houghton Mifflin Co., 1996.
Includes index.
ISBN 0-395-87740-7
1. Science — Miscellanea. I. Title.
Q173.T75   1997
500 — dc21   97-36902 CIP

Printed in the United States of America

Book design by Robert Overholtzer

QUM 10 9 8 7 6 5 4

TO MY OLD FRIENDS
JEFF AND VICKI
AND MY NEW FRIENDS
JARED AND SETH

Contents

3. Astronomy and Cosmology

 4. Earth and Planetary Sciences

5. Biology (Mostly Molecular)

6. Medicine (Mostly Molecular)

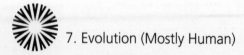 7. Evolution (Mostly Human)

8. Technology (Mostly Computers)

Introduction

IN THIS BOOK I want to look at the enormously diverse enterprise we call science and tell you about the unanswered questions that will drive it into the next century. These are the issues you will be reading about in newspaper headlines over the next decade, the questions whose answers will shape public policy and affect your life in countless ways. The type of car you will drive, the sorts of conversations you will have with your doctor, and the kinds of food you will eat — these are just a few of the things that will be influenced by the progress being made in finding answers to these questions.

One of the abiding sins of science writers is an obsession with completeness — an urge to go back to first principles to explain anything. As a result, the reader often has to wade through page after page of background before getting to the core of the subject. I am as guilty of this failing as anyone, so in this book I've set myself a specific standard: each of the great scientific issues is discussed in no more than three pages. Want to know the most current theories on the causes of cancer? Turn to page 250, and in three pages you'll get a quick overview. Interested in electric car technology? Turn to page 297 for the same.

One advantage of this format is that you can get specific information quickly, without having to sort through a lot of background material. It also means that you don't have to read the items in any particular order. In fact, *this book is NOT meant to be read from the beginning to the end*. It is meant to be browsed in. Each essay can stand by itself. You can read a single one to get the background you need to appreciate a news story. You can use the Table of Connections at the back of the book to find connected topics scattered through the book. Incidentally, if you do read this book from cover to cover, you will occasionally find that in order to keep each essay self-contained, I have had to repeat some information.

Of course, this format has some disadvantages. In three pages, I can't go into much depth on any subject. You won't find much historical context here, and no extended discussions of the philosophical consequences of scientific discoveries. That's all right, though — my colleagues have produced an enormous constellation of books that deal with such subjects.

What are the major unanswered questions in science these days, and how does one go about identifying them? To individual scientists, of course, the subject of their own research looms large, and it is difficult for them to say that any other subject is more important than theirs. Nevertheless, to paraphrase George Orwell, while all ideas may be equal, some are more equal than others. It's a plain fact of life, reflected in a thousand different ways in the world of science, that some questions are more important than others.

When I started to think about this issue, my first impulse was to list all the major scientific questions in order of importance, from 1 to 101. This notion broke down, however, when I asked myself whether I could really make a rational distinction between items 77 and 82 in such a list. Conversations with several friends and colleagues convinced me that although working scientists may agree in general about the most important outstanding questions, consensus vanishes when you get past the top few. So I decided to make a "Top Ten" list of questions, then group the rest according to subject.

To some extent, the choice of items for inclusion in a book like this is subjective and arbitrary, as is the choice of items for inclusion in the Top Ten list. I would argue, however, that these choices are much less subjective than you might think. The reason has to do with the structure of the sciences. Scientific knowledge is analogous to the trunk of a tree, in which a central core of solid wood is surrounded by a thin layer of growing tissue. The basic scientific principles — Newton's laws of motion, natural selection, and so on — are within the core and are unlikely to change. But around that core, scientists are working to add to our knowledge, and that's where the answers to questions in this book are being generated. And just as you can predict from today's tree trunk where next year's growth will occur, so too you can make a pretty shrewd guess as to where next year's scientific breakthroughs are going to occur by looking at today's research efforts.

By and large, then, the questions discussed in this book are the result of looking at what's happening in the sciences and picking out areas where progress is being made. In choosing items to include, and in choosing and ranking the Top Ten, I used a number of criteria, including the following (not necessarily in order of importance):

Is the problem venerable? Questions about the origin and structure of the universe, for example, have engaged the human mind for thousands of years. Progress in this area carries an intrinsic cachet, and including such questions needs no justification.

Is a solution (or significant progress) likely in the near future? I gave less weight to questions that are not likely to be answered within the next decade. This is why one of the Top Ten questions concerns the chemical origins of life, while questions about life on other planets are discussed later, in the sections on astronomy and on planetary sciences.

Will the answer make a difference to us as individuals? Science and technology affect our lives. Many items in the sections on molecular biology, medicine, and technology are included because they are likely to have important effects on us.

Is this topic in the news? Some of the subjects that are most important to scientists are not of particular interest to the public, and vice versa. The search for extraterrestrial intelligence and theories of time travel, for example, don't rate very high on most scientists' lists of major unsolved problems, yet they are of enormous interest to the general public and often appear in the news. On a more serious level, we are much more interested in the evolution of human beings than in the evolution of other species, even though there is no real scientific justification for this bias.

Does the question have implications for the way we do science? Occasionally new advances in one area affect a wide assortment of sciences, sometimes with unexpected consequences. This is why I have included questions related to computer simulation and the solvability of mathematical problems.

Is the question representative of a wider class of issues? Many areas of science don't lend themselves to the question-and-short-answer format, often because the research covers a wide variety of related problems rather than one sharply focused issue. In such a case I take one question as a surrogate for the entire field. The discussion of the hypothetical face on Mars, for example, is included to represent all the fringe areas of science; the story of the planet Pluto stands for all the small and intriguing questions about our solar system.

Does the question have public policy implications? Our understanding of the world shapes how we behave. Thus I have included environmental questions such as those involving possible greenhouse warming and the cleanup of toxic waste, even though these may not be the most important issues from a purely scientific point of view.

Is it fun? I'm only human. Sometimes a topic just strikes me as particularly interesting and neat. In these cases, often against the wise counsel of editors and friends, I've included it in the book. If you don't share my enthusiasm for killer bees (page 229), you can go on to another question. Think of these oddball entries as my reward to myself for doing the work needed to get the other, more solid, items together.

Even with these guidelines, the final choice of whether or not to include a particular question came down to a personal judgment. I think most scientists will agree that I've hit all the high points of our craft, although most would want to make a few adjustments — put a different question in here, remove another one there, and so on. That's what makes the world go around.

Many of the topics in this book are controversial, to say the least. On some I have definite opinions, either about the direction they'll take or about policies that should follow from the science we know. I've made an effort to identify those opinions and keep them separate from the information about mainstream scientific thinking. You are free, of course, to agree or disagree with me, but I think it's important for people writing about science to make the distinction between what is known and what the author thinks. All too often, writers try to bolster their personal views by giving them the patina of scientific truth.

Before turning you loose on the frontiers of science, let me pass on the one surprise I encountered in writing this book. It was not that there are so many unanswered questions about the world — any working scientist could reel off dozens of such questions. The surprise was that as a physicist I found the most interesting and compelling frontiers of science today not in physics or even in cosmology but in biology.

To understand my surprise, you have to realize that for the last several hundred years, it has been physical scientists — physicists, chemists, and astronomers — who have led the way in pushing back the frontiers of knowledge. The results of their work have been truly impressive; among other things, they have produced commercial electricity, the digital computer, and the millions of synthetic materials and pharmaceuticals we use daily. This situation has bred, I suppose, a sense that the physical sciences would always be in the lead.

I was, of course, aware of the revolution going on in biology, triggered by our new understanding of the functioning of living systems at the molecular level and symbolized by the double helix of DNA. What I wasn't prepared for was the depth and breadth of this new brand of biology and the breathtaking speed with which new discoveries are being put to practical use, nor was I prepared for the impact that this work is having on fields outside of molecular biology. Within the next few years, we are going to be deluged with new discoveries about how living systems, including the human body and mind, work, with consequences that are too broad to envision.

I have tried to give some sense of this coming revolution in the pages that follow. Half of my Top Ten questions involve molecular biology in one way or another, as do most of the entries in biology and medicine. At the same time, molecular evidence is playing a bigger and bigger role in the effort to work out the recent evolution of the human race, an area that was once the exclusive domain of paleontologists, anthropologists, and linguists.

Finally, with the usual statement that whatever errors remain in the book are my sole responsibility, I would like to thank my friends and colleagues Harold Morowitz and Jeff Newmeyer for critical readings of the manuscript and invaluable insights and comments.

So with that introduction, flip the book open to a random page and start your tour of the frontiers of scientific knowledge!

James Trefil
Red Lodge, Montana

1

The Top Ten
Problems in Science

1. Why Is There Something Instead of Nothing?

 WHY IS THERE a universe at all? How could everything we see around us have been created out of nothing?

When you begin to think about how the universe began, you naturally wonder what was here *before* it began. The obvious answer is "nothing." But what, exactly, is "nothing"? The best way to characterize current thinking on this question is to say that "nothing" just ain't what it used to be. For most of recorded history, people have had a problem thinking about nothingness, or the vacuum — indeed, recognition of the very existence of such a state is fairly recent. The reason for this difficulty isn't hard to find. Have you ever tried to picture nothing? I can't do it. I can picture empty space surrounding something (two basketballs, for example), but I can't picture the absence of everything. And this shortcoming of human imagination has influenced our thoughts about nature — scientists accepted the existence of the vacuum only when the results of repeated experiments drove them to do so.

But that acceptance didn't last long. With the advent of quantum mechanics, our picture of nothing changed again. Instead of a passive, inert absence of matter, quantum theory tells us that a vacuum is both active and dynamic. According to the laws of quantum mechanics, a bit of matter can appear spontaneously out of nothing, provided that (1) a corresponding bit of antimatter appears at the same time and that (2) the matter and antimatter come together and annihilate each other (disappear back into the vacuum) in a time so short that their presence cannot be directly measured. This process is called the creation of a "virtual" pair of particles, one of matter and one of antimatter.

Think of the vacuum as a level field and the creation of a virtual pair as like digging a hole and piling the dirt up. Then you have a particle (the pile of dirt) and an antiparticle (the hole), but when you put all the dirt back in the hole, you're back to the level field again.

So the modern vacuum is a little like popcorn popping, except that this popcorn can "unpop" as well. A virtual pair pops up here and un-pops, then another pops up there, and so on. And lest you think this is all a fairy tale, I should point out that occasionally a particle traveling through space, such as an electron, comes near one of these virtual pairs and is very subtly altered by the encounter. That subtle alteration can be detected, so the concept of the quantum mechanical vacuum is backed up by more than just imagination!

So the "nothing" from which the universe sprang was not just the absence of everything but a nothing with virtual pairs of very energetic particles popping up and disappearing all over the place. Exactly *how* this sort of vacuum led to the universe we live in remains the big question, and all sorts of theoretical speculations have been advanced about how the system might work. Let me talk about my favorite type of theory to give you a sense of how these theories operate.

Think of the fabric of space as being something like the membrane of a very special kind of balloon. The presence of any matter, even virtual pairs of particles, causes the fabric to bulge, and this drains energy from the gravitational field to make matter. If the bending is severe enough, the balloon starts to expand. In this scheme, if the virtual pairs pop for a long enough time, eventually enough of them will pop in the same place at the same time to bend the fabric enough to start the expansion going. This is the event we usually refer to as the

Big Bang. Oddly enough, calculations indicate that it really doesn't take very much mass to set this process off — about ten pounds packed into a volume smaller than a proton would do the job nicely. In most theories, the energy needed to create the rest of the mass of the universe came from the warping of gravitational fields later on.

This particular version of creation has several interesting aspects. For example, it leaves open the possibility that the process could still be going on and thus that there might be other universes out there. Furthermore, it raises the possibility that we might be able to create our own universes by manipulating matter — what cosmologist Alan Guth calls "the Universe in Your Basement Kit." And finally, it provides writers with some tremendously useful quotes. For example, here's physicist Edward Tyron commenting on the fact that creation may just be a statistical fluke: "Perhaps the universe is just one of those things that happens now and again."

2. Is There a Future for Gene Therapy?

MOST OF MODERN medicine consists of treating symptoms until the body can heal itself. In fact, in only three areas have medical advances given us the ability to eliminate disease — the development of public health and sanitation, the invention of aseptic surgery with anesthetics, and the discovery of antibiotics. Today we may be at the beginning of a fourth development using the new techniques of gene therapy.

In this technique a new gene is inserted into a functioning body cell to correct an error in that cell's DNA or to provide the cell with a new function. The first human gene therapy was done at the National Institutes of Health in 1990. First, blood was drawn from patients who had an extremely rare genetic inability to produce a certain protein. A gene was inserted into the drawn white blood cells, which were then allowed to divide for a few days before being reinjected into the patient. With the added genes, the new cells started to produce the missing protein, and the patients experienced complete cures.

A word about "injecting" genes into cells. This may seem like a difficult procedure, but in fact a certain class of viruses does it all the time. Called "retroviruses," they contain a few strands of RNA and the enzymes needed to turn that RNA into genes (stretches of DNA) and insert those genes into the DNA of a host cell. Much of the work in gene therapy involves understanding how retroviruses do their job and adapting them to the needs of medical practice.

Since that first trial, almost a hundred different genetic therapy protocols have been approved for use in the United

States. And while rare diseases may have been appropriate for the "proof of concept" studies, the first payoff for gene therapy will be in diseases that are much more common, such as cancer, cystic fibrosis, and various blood and pulmonary diseases. In fact, the first commercial application may be a treatment for certain types of brain tumor. In such a treatment (for which a number of preliminary studies have already been done) a retrovirus is injected through a needle into an inoperable brain tumor. (A small hole is drilled in the skull for the needle to pass through.) The retrovirus in this case contains a "suicide gene" — a gene that will, when activated, produce a protein that poisons and kills the host cell. I've seen some "before and after" X-rays of this therapy, and it is little short of miraculous — the tumor just seems to disappear. Similarly hopeful results have been obtained for patients suffering from cystic fibrosis, a disease caused by the inability of cells to produce proteins that move chlorine through the cell membrane. In this case the appropriate retrovirus will be put into the lungs through a tube.

But the real future of gene therapy doesn't lie with procedures such as blood infusions or holes bored in skulls. It lies in making a virus that can be inserted into the body through an ordinary injection (much the way a vaccine is injected) and that will then (1) seek out the target cells on its own, (2) be recognized by those cells, (3) enter the cells, and (4) insert its cargo of genes at a spot on the host's DNA where vital cell functions will not be disrupted. This may sound like a tall order, but natural viruses do it all the time — hepatitis B, for example, seems to have no trouble in finding our liver cells and invading them. A retrovirus that can do all this and not cause disease would be called a "therapeutic virus," and considerable progress has been made in manufacturing them.

All cells have molecules called receptors in their outer membranes. Each receptor has a complex shape that fits ("recognizes") a particular molecule in the environment. Viruses enter cells by presenting shapes that the receptors recognize. Scientists have succeeded in producing viruses whose outer coatings have the right shapes to be recognized by specific cells in the human body. If injected into the bloodstream, these viruses will circulate until they find the right target cells, then bind to their surface.

At the moment, that's as far as we've gone. The process by which viruses actually enter cells and insert their genes is still being studied. When we understand how it works (and I expect that will happen soon), we'll be able to make therapeutic viruses that can be routinely injected in your doctor's office. In 1995, a panel of experts recommended that research of this type, rather than the development of clinical procedures, should be the focus of work in gene therapy.

Today if you have a bacterial infection, you expect to be able to go to your doctor and get a pill or a shot that will clear it up. Can you imagine doing the same thing if you had cancer? That's exactly the promise of gene therapy. After all, we get sick because somewhere in our bodies some molecules are not functioning properly. Gene therapy promises the ability to fix such problems, molecule by molecule.

3. Will We Ever Understand Consciousness?

 "I THINK, therefore I am." With these words, the French philosopher René Descartes (1596–1650) set the stage for one of the great debates of modern times — a debate that is no closer to resolution now than it was in his lifetime. This is the debate over what, exactly, it means for a human being to be conscious, to think, to feel emotions, to have a subjective experience of the world. I include it on my Top Ten list for two reasons: first, it is the only major question in the sciences that we don't even know how to ask and, second, I see techniques developing and schools of thought forming that lead me to believe that this will become *the* scientific question of the twenty-first century.

In all other areas of science, we have a sense of the basic concepts and categories, which allow us to pose the questions in useful terms (by which I mean terms that suggest how to go about finding an answer). In cosmology, for example, we may not have a fully unified Theory of Everything, but we have a pretty good sense of what one would look like and how to go about finding it.

In the area of mind/brain/consciousness, however, we really don't know what the categories are, what the important questions are, or how to go about asking them. Should we be concentrating on the workings of cells in the brain? On large-scale brain functions? On deeper metaphysical or philosophical questions? Right now scientists are groping around trying to sort out these issues. There are numerous approaches to the problem, but my sense is that the researchers are starting to shake out into three broad groups, which I will call the neurophysiologists, the quantum mechanics, and the mystics.

The best known of the neurophysiologists is Francis Crick, Nobel laureate and codiscoverer of the structure of DNA. He argues that the way to understand consciousness is to look at single neurons or collections of neurons in the brain, that all our subjective experience is nothing more than "the behavior of a vast assembly of nerve cells and their associated molecules." Your feelings of joy and sorrow, in other words, are nothing more than the firing of billions of neurons in your brain. People who follow this track study the details of brain functioning (particularly of vision) and try to understand the rich panorama of human experience in those terms. Although you might think that as a physicist I would favor a reductionist approach, I have to say that I hope that these guys are wrong, although in my dark moments I think they may be right.

The quantum mechanics, whose most prominent spokesman is the English mathematician Roger Penrose, argue that the laws of physics underlying ordinary electrical circuits (and the brain as pictured by neurophysiologists) fail to capture the full unpredictability and nonlinearity of the brain. In their view, we won't understand the brain until we have a fundamentally new understanding of the behavior of matter at the atomic level. It is here, they argue, that the origin of consciousness and feelings must be sought, for this last great gap in our understanding of the universe lies at the boundary between the large-scale world (ruled by Newtonian physics) and the small-scale world of the atom (ruled by quantum mechanics). If we can fill in this gap, they argue, we will eventually have a true theory of the mind. Penrose, in fact, argues that this understanding will come from a quantum theory of gravity.

When I think about the third group — the mystics — I picture a stereotypical scene from a 1950s grade-B science fiction movie (a genre to which I must confess a mild addiction). There is often a white-haired, pipe-smoking scientist in these movies, sort of an Albert Einstein clone, who at some crucial moment delivers himself of the opinion: "There are some things, my boy, that science was never meant to know." In the same way, some people (mainly philosophers) argue that human beings either will not, should not, or cannot gain an understanding of consciousness. In some cases the arguments are based on old arguments about the impossibility of deriving a purely mental state from a purely physical system. Other arguments, based on analogies to evolution, state that because humans have never had to understand consciousness, they don't have the mental equipment to do so. But all of them come to the same conclusion: the methods of science alone can never solve this problem.

So how will this question be answered? My own guess is that consciousness will turn out to be an emergent property of complex systems. I suspect we will discover that in the process of hooking up lots of neurons to make a brain, there comes a point where "more" becomes "different." And while this point of view could be accommodated within either the neurophysiological or the quantum mechanical framework, my hope is that when all is said and done, human beings will be found to be something more than "a vast assembly of nerve cells and their associated molecules."

4. Why Do We Age?

 WITH EACH passing year, this question moves higher up on my list of important issues, and if it's not high on your list right now, I guarantee that it will be someday. We all grow older, and the chronology of aging is well known to each of us. But *why* do we age? What mechanism drives the aging process, and is there a way to control it in some way?

A lot of research is going on in this field, and there are probably as many theories of aging as there are researchers. The theories, however, tend to fall into two broad classes — call them "planned obsolescence" and "accumulated accident" theories.

Both types rest upon a simple fact about the evolution of living things: natural selection operates on genetic differences that are passed from parents to offspring. Genetic variations that improve the offspring's chances of living long enough to reproduce are more likely to be passed on, and, over long periods of time, they will spread throughout a population. A condition that comes into play only after an organism has reproduced (and normal aging falls into this category) has no bearing on the production of offspring and therefore does not encounter any evolutionary pressures. As a middle-aged scientist I take little comfort from the fact that in nature's view I might be redundant, but that's the way it is.

"Planned obsolescence" was a term used in the 1950s and 1960s to describe products designed to have a limited life span. Cars, for example, were supposedly built to last less than ten years so that a continuing stream of replacement cars would be needed in future years. (At least that was what consumers believed in those days.) Planned-obsolescence theories of aging suggest that the human body is built so that it will give out after its useful life is over, to be replaced by a newer model.

An important piece of evidence for these theories is the Hayflick limit, named after biologist Leonard Hayflick. In 1961 Hayflick and his colleagues announced the results of a crucial experiment. Cells from a human embryo were put into cultures that provided all the nourishment they needed and protected them from all harmful effects. Given these ideal conditions, the cells started to grow and divide. When a cell got to about fifty divisions, however, the process just stopped, almost as if someone had thrown a switch.

Planned-obsolescence theories hold that the genes in each cell contain a mechanism that turn it off when its time is up. Critics of the theory, while not disputing the existence of the Hayflick limit, argue that most organisms die from other causes long before the limit is reached.

Accumulated-accident theories begin by noting that every living cell is a complex chemical factory containing all sorts of machinery to carry out its functions. That machinery is constantly bombarded by chemicals from the environment — not just the pesticides and pollutants we're used to hearing about but byproducts of the very chemical processes that the cell has to carry out to maintain life. Over long periods of time, according to these theories, the cell's defense mechanisms simply wear out. Aging (and eventual death) are the results. At the moment, evidence seems to be accumulating in favor of this view.

Research on aging in humans is presently focused on understanding the chemical reactions that contribute to the aging process and the kinds of genetic defenses we have against them. The most likely villains now appear to be a group of chemicals known as free radicals, which are a normal byproduct of basic metabolism (as well as of other processes). These chemicals, once free in the cell, break down molecules needed for cell repair and, in some cases, DNA itself. A striking bit of evidence pointing to free radicals as the mechanism of aging is that animals with high metabolism rates (and hence high rates of free-radical production) tend to have shorter lives and to produce fewer chemicals to combat the free radicals.

The hope is that once we understand how these chemicals operate in our cells, we will be able to slow the process of aging. This doesn't necessarily mean that our maximum age will increase — we know almost nothing about why the upper limit of the human life span seems to be about 110 years. Work on the question of whether this limit can be exceeded is just beginning. In the meantime, current research is aimed at making it possible to put off the degeneration of aging, and to remain vigorous later into life.

Come to think of it, the prospect of keeping on going until you die with your boots on isn't so bad, is it?

5. How Much of Human Behavior Depends on Genes?

Or Nature vs. Nurture, *Tabula Rasa* vs. Original Sin, Predestination vs. Free Will

 THIS QUESTION just doesn't seem to go away. And no wonder — in essence, it asks whether human beings are free to behave as they wish or whether their actions are determined in advance. In its modern incarnation, this venerable debate is couched in terms of molecular genetics. Is our behavior determined by our genes, or is every human being shaped exclusively by his or her environment?

For most of the middle part of this century, Americans had an almost religious faith in the second of these choices. Human beings, we believed, were infinitely perfectible, and if people behaved badly, they did so because their environment was bad. Fix the environment, we believed, and you could produce perfect human beings. Evidence to the contrary was considered misguided at best, heretical at worst.

But times have changed. Just as we have come to understand that many diseases have their origins in our DNA, so too have we come to realize that genes play an important (but by no means exclusive) role in determining our behavior. The evidence that changed the view of the behavioral-science community came in many forms. Extensive studies of animals, from fruit flies to rats, showed clear genetic influences on behaviors such as learning and mating. More directly, the large literature of sophisticated studies of twins clearly demonstrate the importance of genetic factors in mental disease and a range of behavioral traits from the vocational interests of adolescents to general cognitive ability. Such studies generally look at either identical twins, who have identical DNA — particularly identical twins raised in different environments — or at fraternal twins, who have different DNA but very similar environments. The general conclusion, based on hundreds of studies, is that genetic factors account for anything from 30 to 70 percent of many human behaviorial traits. In many cases the percentages are higher than those for genetic causation of physical ailments.

In fact, over the past few years the debate has shifted from "Do genes have an influence on behavior?" to "Which gene or genes influence which kind of behavior?" And here there is a great deal of ferment and debate, because the picture isn't as simple as people thought it would be.

For physical diseases (cystic fibrosis, for example) it is often possible to identify a single gene on a single chromosome as the cause of the disease. The original expectation was that behavioral traits would follow the same pattern — that there would be an "alcoholism gene" or an "aggressive behavior gene." In fact, a number of highly publicized claims for these sorts of results were made, but they now are clouded in controversy.

In general, the one gene–one behavior model seems to hold for behaviors for which we can make a clean distinction between members of the population who display it and those who do not. Profound mental retardation and autism, for example, seem to fall into this class. But with behaviors that are less clear-cut, like alcoholism and manic-depressive (bipolar) disorders, initial claims for finding a single responsible gene have been questioned. It may be that more than one gene is involved, or it may be that a complex interaction takes place between genes and the environment. In the case of alcoholism, the complexity may arise because different forms of the disease have different genetic bases. New techniques developed in studies of inbred rats give us hope that before too long we will be able to sort out genetic and environmental influences in these more complex situations.

This area has been and will continue to be enormously influenced by the rapid advances in molecular biology. With the human genome being mapped with greater and greater precision, it has become possible for researchers to scan all twenty-three chromosomes in every subject. Instead of looking for single genes, as in the past, researchers in the future will be casting a much wider net, sorting through huge amounts of data to find families of genes that differ, in people who exhibit a particular behavior, from those of the general population.

As these sorts of data accumulate, we will have a better understanding of the role that inheritance plays in human behavior. I don't expect, however, that we'll ever get to the point of thinking that a person's future is determined entirely by his or her genes. Instead, we'll have a more realistic view of why human beings behave the way they do, a view that will mix environment and genetics in complex and unexpected ways. And I am enough of an optimist to believe that when we are able to discard the outworn "either-or" notions of the past, we'll be well on our way to helping people whose behavior we simply don't understand today.

6. How Did Life Begin?

 FOUR AND A HALF billion years ago the earth was a hot ball of molten rock. Today there is no spot on its surface where you can't find evidence of life. How did we get from there to here?

We already know a large part of the answer to this question. Anyone who has been to a natural history museum probably remembers the dinosaur fossils — exact replicas in stone of the bones of now extinct monsters. In fact, fossils have been formed in abundance over the past 600 million years, preserving the life stories not only of dinosaurs but of many other life forms as well. Before this period, animals had no bones or other hard parts from which fossils could form, but impressions of complex life forms (think of them as jellyfish) have been found going back several hundred million years more. And from the time before that, believe it or not, pretty clear evidence for the existence of single-celled organisms has been found.

It may come as a surprise to learn that we can detect fossil evidence of organisms as microscopic as bacteria, but paleontologists have been doing just that for some time. The technique works like this: you find a rock formed from the ooze on an ocean bottom long ago, cut it into slices, and examine the slices under an ordinary microscope. If you're lucky (and highly skilled) you will find impressions left by long-dead cells. Using precisely this technique, scientists have been able to follow the trail of life back to fairly complex colonies of blue-green algae some 3.5 billion years old. Life on earth must have started well before that.

We also know that the early stages of the solar system coincided with a massive rain of cosmic debris on the newly formed planets. Astronomers call this period the "Great Bombardment," and it lasted for something like the first half billion years of the earth's history. Had life formed during this period, any massive impact would have wiped it out (it would, for example, take an asteroid only the size of Ohio to bring in enough energy to boil all the water in the oceans). Thus we are coming to realize that there is a narrow window in time — perhaps 500 million years — during which life not only must have arisen from inanimate matter but developed into a fairly complex ecosystem of algae.

Scientists have known since the 1950s that it is possible, by adding energy in the form of heat or electrical discharges to materials believed to have been present in the earth's early atmosphere, to produce molecules of the type found in living systems. They also know that in a relatively short time (geologically speaking) such processes would turn the world's oceans into a stew of energy-rich molecules — so-called Primordial Soup. Research today centers on trying to understand how this situation gave rise to a self-contained, reproducing cell.

Here are a few of the current ideas about how this could have happened:

- RNA World — RNA, a close cousin of the more familiar DNA, plays a crucial role in the chemical machinery of cells. Recently scientists have been able to create short stretches of RNA that, placed in the right sort of nutrient broth, can copy itself. Was this the first step on the road to life?

- Clay World — certain kinds of clays have static electrical charges on their surfaces. These charges may have attracted molecules from the soup and bound them together. Once the molecules were formed, they would float away, carrying only indirect evidence of their origin.

- Primordial Oil Slick — my favorite. The same chemical reactions that created the Primordial Soup would have made the kinds of molecules that form droplets of fat in a pot of soup. Each droplet would enclose a different mix of chemicals, and each would be, in effect, a separate experiment in the formation of life. In this theory, the first globule whose chemicals were able to replicate themselves grew, split, and became the ancestor of us all.

Whichever of these (or other) theoretical processes actually occurred, it had to produce that first living cell within a few hundred million years. And this fact, in turn, gives rise to the most intriguing idea of all. If primitive life is really that easy to produce, sooner or later (and I suspect it will be sooner) someone is going to reproduce the process in the laboratory. The product of such an experiment won't look impressive compared to even the simplest cells today, although it will have the same basic biochemistry. Modern cells, after all, have had 4 billion years of natural selection to sharpen their ability to deal with their environments. The experiments will most likely produce a sluggish blob of chemicals surrounded by fat molecules that will take in energy and materials from the environment and reproduce itself. But that blob will forge the last link in the chain between the barren earth of 4 billion years ago and us.

7. Can We Monitor the Living Brain?

EVERYTHING THAT makes you human — your thoughts, your dreams, your creative impulses — comes from a region of the brain that, if it were peeled off and laid out flat, would take up no more space than a large dinner napkin. This is the cerebral cortex, the outer covering of the brain. Today, using a new technique called functional magnetic resonance imaging (fMRI), scientists are learning how this quintessentially human organ works.

It's hard to overestimate the importance of this particular breakthrough. Not so long ago, the only way you could investigate brain function was to inject a radioactive isotope into an animal, kill it, and section the brain tissue. You could then examine how the isotope had accumulated in various regions and deduce which sections of the brain had been active. Later, using a technique called positron emission tomography (PET), molecules containing a radioactive oxygen isotope were injected into a person's bloodstream. Over a period of several minutes, the oxygen isotope would undergo radioactive decay and emit particles called positrons, which could be detected outside the body. Watching a PET scan, you could see regions of the brain "light up" as they drew more blood for their tasks. More recently, radioactive markers have been attached to molecules that are known to interact only with certain types of receptors in specific cells. The result: a map of the locations in the brain where specific molecules do their work.

Like the more familiar MRI, fMRI has become a commonplace tool in medicine. In both techniques, a material is placed between the poles of a large magnet; the protons that constitute the nuclei of hydrogen atoms in that material will then perform a motion called precession. Think of the proton as a top spinning rapidly on its axis. If, in addition, the axis is describing a lazy circle, we say the proton is precessing. The rate at which the protons precess depends very sensitively on the strength of the magnetic field.

If we pass radio waves through a material whose protons are precessing, the waves whose crests correspond to the frequency of the precession will interact with the protons. Typically, technicians shoot radio waves through tissue at different angles, measure what comes through, then use a computer to construct from the radio patterns a three-dimensional picture of the tissue through which the waves have passed.

In ordinary (or structural) MRI, the strength of the interaction is used to gauge the number of protons at various points, and from this information detailed maps of various organs in the body can be made. Functional MRI works in much the same way, except that it is sensitive enough to measure minute changes in the magnetic field at the position of the protons, even those caused by changes in blood flow. These tiny variations in magnetic field reveal which parts of the brain are using more blood when a particular task is performed.

To watch a living brain function, then, you need only to instruct a volunteer to lie down in an MRI machine and perform a mental task, such as thinking of a certain word. The computer constructs a picture of the brain in which the regions that are getting more blood "light up." State-of-the-art machines can differentiate regions to an accuracy of a square millimeter (a millimeter is about the thickness of a dime). This fine resolution is important, because the brain is a highly specialized organ, with each region — there are about a thousand in all — performing specific tasks. For example, if you close your eyes for a moment and then open them again, the seemingly effortless recreation of the visual image of your surroundings involves neurons in dozens of different areas of the brain. Some regions are involved only in peripheral vision, others in seeing the center of the visual field, and so on.

The insights we can get from this new technique go to the very heart of mental functioning. In one study, for example, bilingual people were given tasks that required them to think back and forth between their two languages. It was found that the different languages do not use different regions of the brain. This seems to bear out the current theory among linguists that the brain is "hard-wired" to do language and that different languages are like different programs that run on the same hardware.

The ultimate goal of the fMRI and PET programs is to produce a map of the entire cerebral cortex, showing which regions and combinations of regions are involved in carrying out which mental tasks. The idea is that when you engage in *any* mental activity, from seeing green to thinking about a walrus in your bathtub, the mappers will be able to tell you exactly which parts of your cortex are involved. Like the project of mapping the human genome, the mapping of the brain is an enterprise so fundamental, so profound, that one can only be filled with awe when contemplating it. How fortunate we are to be living in the time when both maps are being made!

8. Are Viruses Going to Get Us All?

YOU MAY REMEMBER that the medicine practiced in *Star Trek* was pretty advanced. They did brain surgery by putting something on the patient's forehead and made complete diagnoses while the patient lay on a bed surrounded by instruments. For all their virtuosity, however, they had not found a cure for the common cold. The more we learn about the effects of viruses on human beings, the more this little plot twist seems like an astonishingly accurate prediction.

A virus is either the most complicated nonliving thing we know about or the simplest living thing, depending on how you define "living." A virus consists of molecules of DNA or its close cousin RNA surrounded by a coating of proteins. This outer coating fools cells into thinking that the virus belongs inside. Once admitted to the inner workings of a cell, the virus sheds its coat and coopts the cell's chemical machinery to produce more viruses. This goes on until the cell's resources are used up and it dies, while the viruses created go on to invade other cells. Thus, although a virus by itself cannot reproduce as living things do, it can reproduce if the right cell is available.

The virus's lifestyle explains why it is so hard to treat a viral disease. A disease caused by bacterial invasion can be knocked out with antibiotics. Many antibiotics work by attaching themselves to molecules that are important in the construction of a bacterium's outer wall and thereby preventing new cell wall from forming. But a virus *has* no cell wall, and any drug that killed the cell in which the virus was working would kill all the healthy cells in the neighborhood as well. To combat a virus, you have to get inside the chemical machinery of the cell itself, something we're only now learning how to do.

In fact, the best defense we have against viruses today involves vaccines that mobilize the body's immune system. With such vaccines we have eradicated smallpox worldwide and eliminated viral diseases like polio as major health concerns in this country. The spread of AIDS, however, is a deadly reminder of what a viral disease can do in the absence of an effective vaccine.

Two things about viruses make them particularly lethal enemies: their mutation rate and their ability to transfer nucleic acids from one virus to another. When cells divide in your body, complex "proofreading" mechanisms operate to make sure that the copied DNA is the same as the original at accuracy levels of better than one in a billion bases. In viruses, on the other hand, there is no such proofreading. Measurements indicate that a single copying of DNA in the virus responsible for AIDS, for example, may contain mistakes at the rate of one per 2,000 bases. This means that viruses mutate at a rate unheard of in cellular life forms. Furthermore, if two or more viruses attack the same cell, they can swap bits of DNA or RNA, producing a totally new viral species in the process.

These two effects acting together mean that the human immune system is constantly facing new kinds of viruses. This is one reason why you have to be inoculated against new forms of influenza virus every year. It also is one reason why a virus that was previously confined to monkeys, for example, can suddenly shift over and start infecting humans.

In the days before widespread travel, a particularly deadly virus could devastate or even wipe out the population in a small area — think of the 1995 outbreak of Ebola virus in Zaire. Such unfortunate disasters did provide a certain kind of protection for the human race, however, because the last virus would expire with the last host. With today's transportation system, however, a new virus can spread to every continent on the earth before we even know it exists. Furthermore, as humans push back the last boundaries of the wilderness, we encounter more and more viruses with which we have had no previous contact, giving them a chance to acquire a new host. Against such viruses, humans have no immediate defenses. It is believed that the original AIDS virus, for example, was a mutated form of a virus that had previously affected only monkeys. All that was needed was for one hunter cutting his finger while skinning a monkey with the mutated virus to loose a deadly disease on the entire world.

Want to think about a real nightmare? Imagine a virus like the one that causes AIDS, which leaves the host alive and able to spread the disease for years before he or she dies. Then imagine that the virus could spread through the air, like influenza and the common cold. How many billions would die before we could deal with it?

In the words of Nobel laureate Joshua Lederberg, "Our only real competition for the dominance of the planet remains the viruses. The survival of humanity is not preordained."

9. When Will We Have Designer Drugs?

 WHENEVER YOU TAKE a medicine, whether it is an aspirin tablet or a high-tech wonder drug, you are implicitly relying on two facts about the universe. The first fact is that life, from amoebas to humans, is based on chemistry. Every cell in your body can be thought of as a complex chemical refinery, with thousands of different kinds of molecules zipping around carrying out their missions.

The second fact about living systems is that the workings of all of these molecules depends on their three-dimensional geometry and a few other simple features. For two molecules to combine to form a third, for example, they must be shaped so that the atoms on their surfaces can form bonds to lock the molecules together. The function of a molecule in a living system depends on its shape. And this is where drugs come in, because the function of the molecules in any drug you take is to alter the shapes of molecules in your cells and thereby alter their function.

Here's an example. The functioning of a bottle is determined by its geometry. It has a large enclosed space for the storage of liquids and an open neck so that it can take them in and release them. If you put a cork in the bottle, it can no longer perform its function, because the cork keeps liquids from entering or leaving the bottle. Using this analogy, a molecule in your cell is the bottle and a drug you take to block its action is the cork.

Historically, people have looked in all sorts of places for molecules that can play the role of medicines. Nature, after all, has been cooking up chemicals for billions of years, so there is a huge supply of possible "corks" out there. Scientists gather specimens of plants and animals from all over the world and analyze them for interesting-looking chemicals which are then tested to see whether they have particular useful properties. If they do, a long development program (typically costing hundreds of millions of dollars) may be launched to shape the molecules into something that can be a useful medicine. The key point, though, is that historically the process has been a more or less random rummaging through nature's storehouse of molecules to find one that might be useful. It would be like finding a cork for your wine bottle by going through a pile of corks until you found one that fit.

This sort of random search has worked pretty well, especially considering that for most of human history it has been done blind, without any understanding of how the drugs worked or (in our language) what the geometry of the affected molecule was. Over the last few decades, however, as we have acquired more detailed information about the molecular biology of diseases, it has become possible to look for new drugs in an entirely new way. Instead of fumbling blindly through a pile of corks to find the right fit for the bottle, we can measure the bottle and make a cork to fit it.

Although a few so-called designer drugs have been on the market for twenty years, I sense that we are at the cusp of a major change and that the pharmaceutical industry is poised for a major advance in this area. We will see lots more of these drugs in the future. (The phrase "designer drugs" is used by scientists to describe molecules made in laboratories rather than taken from nature.)

The primary tool for the development of designer drugs is called Computer Assisted Drug Design (CADD). In this system target molecules are shown on a computer screen; then various candidate drug molecules can be brought up to the target to see if they fit. In effect, the computer is used to see if the "cork" fits into the "bottle." This step replaces the laborious process of collection and laboratory testing used to find the raw materials for drugs today.

If the particular design of a drug molecule seems promising, then other questions can be asked — is it the most efficient design for the job? Can it be made in the laboratory for testing? Can it be manufactured in large quantities? Cheaply? Will it be absorbed by the body so that it actually gets to the site where it is supposed to work? And finally, will there be harmful side effects? Some of these questions can be answered at the CADD stage, but others (particularly the last) have to be answered through clinical trials.

Oddly enough, the major effect of designer drugs may not be in medicine (as important as they will be there) but in our attitude toward the environment. A standard argument for preserving the rain forests is that they have in the past served as reservoirs of molecules that can be used as drugs. But if we can design these drugs from scratch, this argument about the value of rain forests becomes less compelling. Perhaps we ought to think about finding another argument.

10. Is There a Theory of Everything, and Can We Afford to Find It?

THE UNIVERSE is a pretty complex place, yet for millennia scientists have harbored a seemingly crazy thought. "The universe may *look* complicated," they say, "but if only we were smart enough and persistent enough and had enough equipment, we could see that underneath all that complexity is a world of surpassing simplicity and beauty."

Here's a simple analogy: you go into a new town and see all kinds of buildings around you, from skyscrapers to bungalows. When you look at the buildings closely, however, you notice that they are all just different arrangements of a few kinds of bricks. Apparent complexity, underlying simplicity.

The search for underlying simplicity has driven an enormously successful scientific enterprise that, by some reckonings, can be traced back two millennia to the Greek philosophers. In its modern incarnation, it burst on the scene at the beginning of the nineteenth century, when John Dalton suggested that although there are huge numbers of different kinds of materials in the world, they are all made from different combinations of a relatively small number of atoms. In this century we have followed the trail down through the atom to its nucleus, and through the nucleus to the particles that make it up. Most recently our attention has centered on quarks, which we believe are the most fundamental units of matter. In this way of looking at things, when you pick up a pencil and ask "What is this?" you answer by talking about assembling quarks into elementary particles, then assembling the particles into atomic nuclei, then tacking on electrons in orbit to make atoms, and finally, locking the atoms together to make the wood and graphite in the pencil. Apparent complexity, underlying simplicity.

And over the past few decades we have started to suspect that that underlying simplicity may be even simpler than the discovery of quarks suggests. It would be as if we had discovered not only that all the buildings were made of bricks but also that if we looked closely enough, all the different fasteners — nails, glue, mortar, and so on — were made of one kind of material. Mortar is one way that constituent parts interact to form larger structures. In the universe the role of mortar is played by the forces that govern the way one bit of matter interacts with another.

On first glance, there appear to be four such forces in nature — the familiar forces of gravity and electromagnetism and, on the atomic scale, the strong force (which holds the nucleus together) and the weak force (which governs some kinds of radioactive decay). What physicists have learned is that if you look at these forces at a high enough magnification, they appear to become identical. Although there appear to be the equivalents of nails, mortar, glue, and staples in the nucleus, if we are clever enough we can see that there is only one kind of fastener. Apparent complexity, underlying simplicity.

Theories that describe the coming together of the four forces at high magnification (or, equivalently, at high energy) are called unified field theories. The first such theory showed that at energies attainable at the biggest particle accelerators on earth, the electromagnetic and weak forces are seen to be the same. The next step, predicted by the so-called Standard Model (our best theory of the universe to date), is the unification of the strong force with these two. The final theory, as yet little more than a gleam in physicists' eyes, would unify all four forces in a Theory of Everything.

Think about it. This would be one equation, probably something that could be written on the back of an envelope, that in a sense would contain everything we can know about the universe.

When I started out in my career as a physicist, I thought there was a reasonable chance that ours would be the generation that finally wrote this equation. I no longer think so. In order to make progress in science, there has to be an orderly interaction between new theories and new experiments. The two perform a kind of waltz through time. At the moment, the process is stalled because although there are plenty of theoretical ideas, there are no relevant data against which to test them. To get such data, we need a machine with enough energy to produce it. This machine was supposed to be the Superconducting Supercollider, a $10 billion machine to be built in a fifty-two-mile circular tunnel under the Texas prairie south of Dallas. Unfortunately, in 1993, with a full fourteen miles of tunnel already dug, Congress stopped funding for the machine. Some future generation, perhaps in a country blessed with a more farsighted government, will complete the dream that began two millennia ago with the Greeks.

And we will be just a footnote in history — the ones who frittered away the time between the first man on the moon and the Theory of Everything.

2

The Physical Sciences

Is There Really a Science of Complexity?

DURING WORLD WAR II, American scientists motivated by the need to guide and control the huge military effort developed a branch of mathematics called the theory of systems. This dealt with issues like finding optimum construction schedules, optimum supply routes, and the like. For a brief period after the war, scientists wondered whether there might be a science of systems — whether something like a transportation web might behave in a certain way simply because it is a system. The thought was that all systems might share certain properties, much as metals conduct electricity by virtue of the fact that they are metals. As it turned out, the answer to this particular question was no. There are many systems, each of which can be analyzed in some detail, but there are no properties or theories of systems in general.

This is not to say that looking for a general theory was totally pointless — important advances in the modern sciences grew out of it. It simply means that a common name does not imply any kind of common behavior.

Today the new darling of the sciences is a subject called complexity, particularly things called complex adaptive systems. And once again the air is thick with speculations about whether there is a general theory of complexity — do widely disparate systems have certain common properties just because they are complex?

The first thing you have to realize is that despite (or maybe because of) all this excitement, scientists do not even agree on a definition of complexity. Indeed, at a recent meeting on the subject I heard someone remark that scientists would sooner use someone else's toothbrush than someone else's definition.

Let me give you a sense of this debate by describing one simple definition of complexity, which has to do with the size of the computer program you would have to write to reproduce the behavior of a complex system. For example, if the "system" were the set of even numbers, you could reproduce it by telling the computer to multiply every number by two. The instructions to the computer would require a lot less information than would be required to write down all the even numbers, so this system wouldn't be very complex. In this scheme, the closer the amount of information needed by the program gets to the amount of data being reproduced, the more complex the system is. This kind of complexity is called "algorithmic complexity" ("algorithm" refers to the set of instructions or rules for solving a problem).

A complex adaptive system is one that has many different parts, each of which can change and interact with all the others, and one that as a whole can respond to its environment. The brain can be thought of as a complex adaptive system, as can economic systems in societies and even some large molecules.

The general tool for studying complex adaptive systems is the computer model. There are all sorts of computer simulations (many bearing an uncanny resemblance to commercial computer games) that can model the ability of complex systems to grow, change, and adapt to different environments. As of this writing, however, no one has been able to generalize from the various computer simulations to general rules for complex adaptive systems — there is not, in other words, a general theory of complexity at this time.

If such a theory is ever found, one feature that I think it will have is something called self-organized criticality. The easiest way to visualize this property is to imagine pouring sand into a pile. For a while the sand just piles up. This is called subcritical behavior. At a certain point, however, the slope of the pile is steep enough that you begin to get small avalanches down the sides. This is called the critical size of the sand pile. If you add sand very slowly, one grain at a time, to achieve a steeper slope than this, the pile is in what's called a supercritical situation. In this case, you will have large avalanches until the slope is reduced to its critical size. The sand, in other words, responds to a changing situation by returning to the critical condition.

This example illustrates many properties that seem to be shared by complex systems. It involves large numbers of "agents" (sand grains), which at a certain level of aggregation exhibit behavior simply because there are many of them. It would be very difficult to predict the action of each grain, but we can estimate how many avalanches will occur in a given time period. Thus there are behaviors of complex adaptive systems that are easy to predict and behaviors that are more difficult.

Aside from the fact that they share this behavior, though, I think we will find that complex systems like sand piles and the human brain have no more in common with each other than simple systems do.

How Much of the World Can We Simulate with Computers?

 TRADITIONALLY, there have been two basic branches of science: theory, which attempts to construct mathematical models of the universe, and experiment / observation, whose task is to test theories and to determine the kind of universe we actually live in. Over the past decade, however, some thinkers have suggested that with the advent of the digital computer a third kind of science has been born — the science or art of computer simulation.

A computer simulation works like this: the laws that govern the behavior of a physical system (a star, for example, or the earth's climate) are reduced to a set of equations fed to a computer, along with the numbers that describe the present state of that system. The computer is then asked to use the equations to tell how the system will look one time step into the future, then repeat the process for the next step, and so on. In this way the evolution of the system can be predicted. And although the process I've described could be applied even to simple systems, the term "simulation" is customarily used to refer to complex systems in which changes in one part affect what happens in other parts.

Let's take the prediction of future climates in the face of our increased production of carbon dioxide (the greenhouse effect) as an example of a computer simulation. These simulations start by breaking the atmosphere into a series of boxes. Each box is several hundred miles wide, and there are about eleven boxes stacked from the bottom to the top of the atmosphere. Inside each box, the familiar laws of physics and chemistry describe what is happening. Taking into account the fact that air, heat, and water vapor can move from one box to another, we can predict how the atmosphere will evolve. When the books have been balanced, the atmosphere displayed by the computer is a simulation of the atmosphere of the earth a short time in the future. If we keep going, the computer will produce a picture of the earth's atmosphere farther and farther ahead. In fact, we often see the results of such calculations reproduced in discussions of the greenhouse effect. The computer models are called "global circulation models" (GCMs).

In fact, the GCMs are a very good example of both the strengths and the weaknesses of this new way of doing science. If you want to predict what will happen if the amount of carbon dioxide in the atmosphere is doubled, it's obviously easier (not to mention safer) to do it in a computer than on the real earth. On the other hand, when you're confronted with dramatic graphics about the world inside the computer, it's easy to forget that that picture is not necessarily the same as the real world it's supposed to simulate. How close it comes depends on how much you know about the system being simulated and how skillfully you manipulate this information in the computer.

To take the GCMs as an example again, there are a number of important features of the real earth that aren't built into the simulations. We know, for example, that clouds play an important role in determining temperature. Some clouds reflect sunlight back into space, others trap heat at the surface. Any realistic description of the climate, then, should contain a realistic description of clouds, yet in the GCMs every patch of sky either contains clouds in a block hundreds of miles wide or is clear. The clouds on the planet in the computer, in other words, are nothing like the clouds on planet Earth.

In the same way, the effect of the oceans and their interactions with the atmosphere are handled poorly. One popular version of the GCM, for example, treats the ocean as if it were a "swamp" — gases are exchanged between air and water, but there are no ocean currents like the Gulf Stream. Again, the ocean in the computer doesn't bear much resemblance to the real ocean. In this case, as with the clouds, the problem is that even the most powerful computers simply can't handle the full complexity of the real earth. We can make detailed models of ocean circulation, but it's hard to put them into the same computer with a GCM and get answers back in a reasonable time.

Finally, there are some aspects of the earth's climate that we don't understand well enough to put into the computer at all. The effects of soils and vegetation on rainfall, the growth and shrinking of sea ice, and the possible variations in the energy from the sun are all examples.

Once you've gone through something like the GCM in detail, you have a healthy skepticism about computer simulations. The first question you should ask when you see the results of a simulation is how well the system in the computer matches the system in the real world. In the lingo of simulators, this is called validation of the model. The best validation is to apply the simulation to a situation where you already know the answer. You could, for example, feed in climate data from one hundred years ago and see if the GCM predicts the present climate. The fact that GCMs can't do this is one reason I take their predictions with a grain of salt.

Where Do We Go Without the SSC?

REMEMBER THE Superconducting Supercollider? The SSC was supposed to be the next step in the quest to discover the ultimate nature of matter, a quest that began 2,000 years ago with the Greeks. In the largest high-tech construction program ever attempted, the SSC was going to occupy a fifty-two-mile circular tunnel under the plains south of Dallas. Unfortunately, Congress cut off funding for the project in 1993, and today the fourteen miles of completed tunnel are slowly filling with water, a monument to America's inability to focus on long-term goals.

Now I don't want you to think I'm bitter about this. The fact that in my darkest hours I think that one of humanity's noblest dreams was cut off by an alliance of political hacks and a few scientists whose level of bad judgment was exceeded only by the depth of their envy really shouldn't sway your thinking one way or the other. Whatever we think, the SSC is dead, and we have to turn to the question of what is going to happen in the foreseeable future.

The search for the ultimate nature of the material world has been a succession of realizations that each type of matter is made up of smaller, more fundamental units. Ordinary materials are made from atoms; atoms contain nuclei, which are themselves made from elementary particles; and these particles are made from quarks, which are more elementary still. As scientists worked their way down through this series of boxes within boxes, they found that at each new level you need more energy to learn about the nature of the particles — and much more money.

It's easy to break up atoms — we do it all the time in fluorescent bulbs. To break up a nucleus, however, requires a million volts and a million dollars. To get at the elementary particles requires a billion volts and tens of millions of dollars. To study the quarks and the first stages of the unified field theories — the current frontier — requires almost a trillion volts and hundreds of millions of dollars. The SSC, which would have been the next step, was slated to supply almost a hundred trillion volts and would have cost about ten billion dollars.

This kind of progression of cost in equipment isn't unique to the study of elementary particles. In 1948, when the great telescope on Mount Palomar in California was finished, it cost about seven million dollars — as much as the most complex highway interchange of the day. In 1992, when the Keck, then the world's largest telescope, was installed on Mauna Kea in Hawaii, it cost a hundred million dollars — still about the same as the largest freeway interchange in the country. Whether you look inward or outward, it seems, research costs more and more.

At any given stage in the development of a science, however, there is some experiment that must be done, some machine that must be built if progress is to continue. The SSC was that machine for high-energy physics in the twenty-first century. It was designed to deal with a number of questions, the most important of which was the nature of mass.

Mass remains one of the most mysterious properties in the universe. We know, for example that the electron has mass, and we can even measure it to several decimal places. What we can't do is tell you *why* it has mass, and why its mass is different from that of a proton. The theories that would have been tested on the SSC suggest that the answer to this sort of question involves an as yet undiscovered particle called the Higgs boson. Named after the Scottish theoretician Peter Higgs, this particle is thought to permeate the universe, like an exotic magnetic field. And just as different magnets interact with the earth's magnetic field with different strengths, the theories predict that different particles will hook up in different ways to the Higgs background. We perceive these interaction strengths as mass, so questions about mass come down in the end to the properties of the Higgs boson. The SSC was designed to produce those Higgs particles in large numbers, which is why it was so important.

So what do we do while we're waiting for another SSC? Some physicists are taking a new look at existing accelerators, trying to substitute cleverness for power and see if they can conduct experiments that will shed some light on the nature of the Higgs. For example, there are proposals to use existing machines to make large numbers of particles containing the rarer quarks, the hope being that with a large number of data points some features will emerge that aren't evident today. Others are pinning their hopes on a new machine to be built at the Center for Nuclear Research (CERN) in Geneva, Switzerland. Called the Large Hadron Collider, or LHC, this machine was originally conceived as a quick and dirty version of the SSC. Lacking the SSC's energy, it will be able to produce the Higgs particle only if the Higgs has the lowest mass that theorists expect. But as the design for this machine goes forward, the inevitable financial pressures within the European community keep lowering the energy it will have and pushing back the completion date — it's at 2008 now. The problem: if the LHC doesn't find the Higgs, does it mean that the theories are wrong or that the machine just didn't have enough energy? We'll never know until we build the SSC.

How Much of the World Is Chaotic?

 ONE OF THE most startling discoveries made by theoretical physicists and mathematicians in the last two decades was that there are systems in nature — sometimes very simple systems — whose behavior cannot, for practical purposes, be predicted. These systems were given the colorful name of "chaotic" and, as often happens these days, the idea of chaos was oversold. Most natural systems are not chaotic, but enough of them are to make the subject interesting. One of the most intriguing questions concerns the boundary line between chaotic and nonchaotic behavior. What's on one side, what's on the other?

First, let's talk a bit about what a chaotic system is and in what sense it is unpredictable. Think of an experiment in which you drop little chips of wood close together into a river. If the river is deep and smoothly flowing, then chips dropped near each other will stay together as they are swept downstream. And if you double the distance between the chips at the point where they're dropped in, they'll be twice as far apart downstream as they would have been had you not increased the distance. This is an example of a linear, predictable system.

Now imagine that instead of a deep river, we drop wood chips into a whitewater rapids. Two chips dropped near each other on the upstream side of the rapids will, in general, be far apart when they get to the other end, and there will be no simple linear relationship between how far apart they were at the beginning and at the end. This is a chaotic system. The slightest change in the initial state produces huge changes in the outcome.

And that's where predictability comes in. If you know with absolute mathematical precision where the wood chip is when it enters the rapids, you can predict exactly where it will come out. If, however, that initial position is even the slightest bit uncertain, the actual downstream position of the chip will be far away from the predicted position. Since in the real world there is always some uncertainty in measuring the initial position of the chip, in practice it is impossible to predict where the chip will come out.

If a system in nature is chaotic, then, it is impossible to predict its future in practice (although it is always possible to do so in principle if you can measure the initial state with total precision). In fact, in a chaotic system, the separation between systems that start out near each other must grow exponentially with time.

Although it may not be obvious, the analysis of whitewater and other chaotic systems has had a major philosophical impact on our thinking about the nature of the physical world. From Newton on, it was assumed that if you could write a mathematical equation describing a system, then you could predict its future. We thought, in other words, that the ability to describe implied the ability to predict. What the discovery of chaos has shown us is that this is not necessarily true — there are systems we can describe with equations but still can't make predictions about.

Since the discovery of chaotic systems, another question has come to dominate discussions among scientists. What systems in nature are chaotic? Some answers to this question are obvious — turbulent water, the stock market, and the weather are almost certainly chaotic. But other answers are surprising. Recently, for example, some scientists have claimed that many features of the solar system — the quintessential example of Newtonian predictability — may be chaotic. Complex computer models that track the paths of the planets and all of the gravitational forces between them seem to show that over hundreds of millions of years planetary orbits may, in fact, be chaotic. These conclusions have come from studies that first predict the orbits of the planets far into the future from one beginning position, then theoretically advance the planets a few inches along in their orbits at the beginning and recalculate. The results seem to indicate just the sort of divergence we talked about for wood chips in whitewater. I have to admit to a little skepticism here — there's no doubt that the computer simulations of the solar system are chaotic. The question is whether those simulations really represent the world we live in.

I don't want to leave you with the impression that the existence of chaos is a totally negative thing for science, however. As always happens with new discoveries, people are already thinking of ways to use this phenomenon. Chaotic systems, for example, are perfectly suited for applications in cryptography. If two people know the equation that describes a chaotic system, they can use that system as the basis for a code. Using the equation, they can send a message anyone else will see as only a series of random signals, but which they will be able to understand. Similarly, some scientists have recently left their academic positions because they think they can use their expertise in chaos studies to understand the stock market. (I'll believe *that* was a wise move when I see those guys driving around in Rolls-Royces and making killings on the market!)

If You Played the Tape Again,
Would You Get the Same Tune?

THINK ABOUT the chain of events that had to take place for you to be sitting where you are reading this book. Out of all the billions of people on the planet, your parents had to meet. Someone (a teacher? a parent?) introduced you to the joys of reading. You happened to walk by a bookstore or library and see the book — the list goes on endlessly. One interesting question that comes from such speculation is this: if we went back to the beginning of the universe and started over, would a new you still wind up sitting there reading a new version of this book?

Although this may sound like a theme for a science fiction novel, in fact it touches on a very fundamental question about the nature of the universe we live in. To what extent is the behavior of the universe governed by fixed, deterministic laws and therefore predictable, and to what extent is it the result of unpredictable events, of chance? How much of the universe is physics, in other words, and how much is history? The answer to this question changes with the philosophical fashion. In the eighteenth century, the weight of opinion would certainly have been on the side of deterministic predictability; in the twentieth, on chance (although the pendulum may be starting to move back the other way).

In two areas of science this question has been taken beyond idle speculation and into the realm of serious debate: cosmology and evolutionary theory. In cosmology, the question is usually framed in terms of the fundamental constants of nature. If the gravitational force, for example, or the charge on the electron were different, what would the universe be like? In particular, would it have been possible for an intelligent being to develop and ask a question like this?

If the gravitational force were significantly weaker than it is, it could not have pulled material together to form stars and planets. If it were significantly stronger, the expansion of the universe would have reversed itself soon after the Big Bang, and the universe would have collapsed before stars or planets came into existence. In either case there would have been no life and no one to ask the question about the force of gravity.

Reasoning like this shows that only a relatively small range of values of phenomena like the gravitational force or electrical charge will allow the possibility of life. In universes outside of this range, the questions can't be asked; the fact that they are being asked in our universe means we're within the range. This sort of argument is often referred to as the "anthropic principle."

On the evolutionary side, the argument centers on whether, and to what extent, living systems *have* to be like the ones we observe on earth. For example, we know that the first reproducing cell was formed from molecules generated from inorganic material. The question: was that event the result of a chance coming together of random molecules or was it in some sense determined by the laws of chemistry and physics?

Not too long ago, there was a consensus for what we could call the "frozen accident" school of thought, which argued that life could have formed in many possible ways and that our own form just happened to get there first. But as we have come to realize just how fast life arose on our planet and how fast prebiotic reactions can take place in the laboratory, this thinking has started to change. Many scientists in this field — perhaps even most — now think that life didn't originate by a series of chance events but was driven by natural laws. One important test of this notion, of course, would be to find the remains of life that originated elsewhere (on Mars, say) and see if it's like us.

Once we get into the chain of evolution that followed that first cell, the arguments become more murky. The major problem here, I think, is that the process of evolution is designed to fit an organism into its environment, and that environment is always changing. Consequently, no example of evolution in the real world can teach us a crisp textbook lesson. For example, about 550 million years ago skeletons and hard parts first appeared, and it became possible to see abundant evidence for life in the fossil record. Many of the forms that can be seen in those early fossils are unfamiliar to us today. From this some paleontologists have argued that the forms that became extinct represent the workings of chance — that there, but for the grace of God, go we. Others, however, have pointed out that at least some of those early forms have descendants walking (or, more correctly, slithering) around today. Besides, they argue, how can you say that the extinctions that occurred were due to chance when you don't know what forces were operating to shape evolution in that early environment? Maybe those life forms just couldn't have survived anywhere.

This argument will go on. Don't expect it to be solved anytime soon, and be very skeptical of anyone who claims to have a definitive resolution of it.

Are There Problems
That Can't Be Solved?

MANY YEARS AGO, I spent a pleasant year at a laboratory in Geneva, Switzerland. Anxious to improve my French, I arranged to "trade languages" with a Swiss electrical engineer. Every day we would meet for lunch and for forty-five minutes speak English. Then we would retire to another room for coffee and forty-five minutes of French. One afternoon, my friend said something I have never forgotten: "The trouble with you, Jim — the trouble with all Americans — is that you think every problem has a solution."

Over the years, as my age and (I hope) wisdom have increased, I have come to accept the idea that some social and political problems, if not insoluble, are close enough to make no difference. I still cringe, though, at the thought that there may be physical or mathematical problems in this class.

Actually, we can talk about two kinds of issues under the heading of insoluble problems. One involves the logical foundations of mathematics itself; the other, modern computing. At the end of the last century, there was a serious discussion among mathematicians about whether or not it was always possible to determine if a mathematical statement was true or false. In 1901, Bertrand Russell discovered a paradox that cast doubt on whether this was possible. (Here's a simple form of Russell's paradox. In a certain town, the barber says he will cut only the hair of those people who don't cut their own. Does he cut his own hair?) Later, in 1931, the Austrian mathematician Kurt Gödel proved a theorem (now named after him) that says that any sufficiently complex mathematical system will include statements that are clearly true but that cannot be proved within the system itself.

In the computer age, however, the discussion no longer centers around whether problems can, in principle, be solved but over how long it would take a computer to solve them. It turns out that there is a hierarchy of complexity among problems, based on how rapidly computing time increases when the size of the problem is increased, assuming that the computer is given the most efficient possible set of instructions for solving that problem.

For example, suppose you feed census data into a computer that then tells you the total population of a certain area. Suppose that when you fed in information on one city block it took the computer a millisecond to get the answer. You might expect that if you fed in data on two city blocks it would take no more than two milliseconds, ten city blocks no more than ten milliseconds, and so on. (In reality, a good program would probably take a lot less time.) Such a problem is said to be "tractable" or in "complexity class P" because the time it takes to solve an expanded problem increases (in this case) linearly — doubling the input does no more than double the time required. In fact, the problem would be considered tractable even if the time increased as the square or cube or some other power of the input (that is, if going to two city blocks increased the time by four, going to three blocks increased it by nine, and so on). Mathematical expression involving these sorts of powers are called polynomials, which explains the "P" in the name of the problem class. A problem that can't be solved in polynomial time is said to be "intractable."

The next step up in complexity is the class of NP (nonpolynomial) problems. An example is the so-called traveling-salesman problem, which asks the computer to lay out an itinerary that sends the salesman to each city he needs to visit once and only once. Obviously, the more cities to be visited, the longer it will take a computer to solve the problem. (This doesn't mean that a specific example of the traveling-salesman problem can't be solved on a computer — for any specific finite number of cities, the problem can be solved if you're willing to wait long enough). NP-class problems have the characteristic that if you guess at a solution, you can verify that it works in polynomial time, but no one knows whether the best possible set of instructions to the computer will run in that sort of time frame or not.

The picture is further complicated by the fact that many NP problems are equivalent to each other — that is, if you know the solution to one, some simple manipulations will give you the solution to the other. There is even a set of NP problems (called "NP complete") that can be proved to be the worst-case scenario as far as complexity goes. If any one of them is tractable, then all NP problems are as well. Similarly, if (as most mathematicians expect) any NP-complete problem is intractable, then they all are. Figuring out how to solve NP-class problems is the modern frontier of tractability theory.

This discussion of tractability isn't purely theoretical. In many cases engineers need to look at problems in finer and finer detail to produce efficient designs, a procedure analogous to increasing the number of cities in the traveling-salesman problem. Knowing that the computer program won't have to run indefinitely is both crucial and practical.

Is a Quantum Theory of Gravity Possible?

IF THE BIGGEST thing you can think of is the entire universe, the smallest might be an electron or a quark. If you're a theoretical physicist, you believe in your heart of hearts that a single law — a single way of describing things — explains them both. Oddly enough, the main obstacle in the quest to find that law is understanding the most familiar of the forces of nature — gravity.

In our everyday world, we think of forces as something simple — a push or a pull. In the early part of this century, Albert Einstein developed the theory of general relativity. This theory remains our best description of the force of gravity and is particularly useful for describing interactions over large distances, as in galaxies, or around very massive objects, such as black holes.

In general relativity, gravity is thought of as arising from the warping of space-time. When a light ray bends as it passes near the sun, we think of the mass of the sun distorting the space around it, more or less the way a heavy weight would distort a stretched awning. Light entering this distorted space curves, and we interpret this effect as being due to the force of gravity. In Einstein's view of the universe there is no force of gravity, only the geometry of a warped space-time continuum.

At the other end of the size scale, the theory that describes the interaction of particles inside the atom is quantum mechanics. When we talk about the electrical force that keeps electrons in orbit around the nucleus, we think of it as coming from the exchange of bundles of mass and energy. Picture two people walking down the street throwing a baseball back and forth. The people would have to stay fairly close together to do this, and an outside observer might see the baseball as generating a force that held the people together. In quantum mechanics, we imagine that particles in the atom lock themselves together by exchanging other particles; the electron orbits the nucleus because the two exchange photons. Similarly, you don't float off into space because a flood of particles called gravitons are being exchanged between you and the earth, generating the force of gravity. Obviously, this view of gravity is very different from that of general relativity.

Normally this difference doesn't matter to scientists, because they work on problems in which only quantum mechanics or only relativity has to be considered. If you wanted to include gravity in working out the electrical force between an electron and a nucleus in the atom, for example, you'd find that it mattered only in the thirty-seventh decimal place! Similarly, what happens inside an individual atom has virtually no effect on the behavior of a galaxy.

But what about the early universe, where all the mass was contained in a volume smaller than an atom? What about black holes, where concentrations of mass create enormous gravitational forces? In these areas, scientists have to confront the fundamental difference between gravity and other forces.

A theory that treated gravity as a force generated by the exchange of particles would be called a quantum theory of gravity, or simply quantum gravity. Although scientists have been trying to develop such a theory for the past half century, we do not yet have one. Not only would a quantum theory of gravity be a major step on the road to a Theory of Everything, but it has been suggested that it might give us new insights into some of the weird properties of quantum mechanics.

Think of the problem of developing quantum gravity as being something like climbing a high mountain. There are two possible strategies. First, you could simply try to climb it all at once. This approach might seem straightforward, but you risk failing to reach your goal. Or you could start by climbing some small hills. You know you can get to the top of these, and you hope that what you learn will help you in a later assault on the mountain.

People who take the first approach to quantum gravity develop complete theories from the start, theories with names like "supersymmetry" or "string theory." These involve enormously difficult mathematics and, to date, no one has been able to work through them. They are gallant attempts to scale the mountain all at once, but like all such ambitious attempts, they are difficult to pull off.

Other people (most notably Steven Hawking) have been trying the second road to the top. Typically, their theories incorporate pieces of both general relativity and quantum mechanics, but only those pieces that give us equations we can actually solve. These attempts have had some success. We have learned, for example, that black holes continuously spew out particles (called Hawking radiation) and, over long periods of time, evaporate away. Presumably, a full theory would give us even more interesting insights into black holes and the origin of the universe.

The Quantum Eraser

LOOK, THERE'S no way I can explain the absolute weirdness of quantum mechanics to you in a couple of pages. Suffice it to say that I firmly believe that with this branch of science devoted to the study of the behavior of matter and energy at the subatomic level, we have reached a fundamental milestone. For the first time, we have encountered an area of the universe that our brains just aren't wired to understand. You can't picture what goes on inside the atom, even though you can predict the outcomes of experiments to dozens of decimal places. For most people (myself included), this is a somewhat disturbing state of affairs. And, I'm sorry to say, some experiments that are on the horizon are going to make it worse.

Some background: in our everyday world, we are accustomed to things being either particles (think baseballs) or waves (think surf at a beach). We are accustomed to things that have definite locations — they're either there or not there. But these intuitions don't hold in the quantum world. The textbook way to explain this is to talk about something called a "double-slit experiment." Think of a wall that has two parallel horizontal slits cut into it. If you let particles hit this wall, you will see some of them accumulate in piles behind the slits. If you let waves impinge on the wall, you will see behind the wall alternate regions of high waves and no waves at all, from waves that came through the two slits and either reinforced or canceled each other out. In our everyday world, then, it should be easy to tell particles from waves — just throw something at a double slit and see what it does. If you get a banded pattern, they're waves. If not, they're particles.

Now let me tell you about what happens in the quantum world. Suppose you shine a beam of electrons on a double slit in such a way that only one electron goes through at a time. Each electron will land at a definite spot on the other side of the slits, just as a baseball would. But if you repeat this operation many times, shooting single electrons at the slits and recording where they land, you will find that the electrons accumulate in bands corresponding to the reinforcement of waves and that no electrons land in those places that correspond to waves canceling each other out.

Don't try to picture how this happens in terms of everyday objects — that way lies madness. Just take my word that the science of quantum mechanics predicts this outcome exactly, without having to worry about how one electron "knows" what other electrons will do when they come through the slits. It has to do with the fact that objects at the quantum level are neither particles nor waves in the classical sense, but something else that has properties of both. In this experiment, electrons exhibit their wave properties by sensing both slits at the same time, while exhibiting their particle properties by registering at a particular point on the other side. But if you put detectors near the slits to find out which slit the electron went through, you find that the act of detecting the electron changes its behavior, and the banded pattern disappears.

Although these experiments were done decades ago, the notion that the quantum world could be so different from ours still rankles. And now a group at Munich is developing a high-tech version of the double-slit experiment that will be even harder to square with our intuition. It will be done with atoms rather than electrons, and in front of each slit will be a laser apparatus and a box. The laser tickles the atom in a way that causes it to rearrange its electrons but does not change the wave/particle properties of the atom as a whole. As a result, the atom leaves some microwave radiation behind in the box to mark its path, then goes through the slit and registers on the other side.

Here are the predicted results from quantum mechanics: if the experiment is done with the lasers turned off, you'll see the kind of banded wave structure described above. If it's done with the lasers on, so that we can say which slit each atom went through, the banded structure will disappear and we'll see a pattern characteristic of particles. In effect, the fact that the atoms left the microwaves as a calling card and we measured them "freezes" each atom into its particle-like state. But here's the zinger. Suppose we do the experiment one atom at a time, but after each atom has gone through, we open a gate between the two boxes so that the microwaves can be in either. Suppose, in other words, we "erase" the record that tells us which slit the atom went through. Then quantum mechanics predicts that we will get the banded structure characteristic of waves even if we open the gate for each atom *after that atom has arrived and been recorded on the other side of the slits!*

This prediction is so wild that some physicists go ballistic when they talk about it ("necromancy" is one of the milder epithets I've heard). But why do we react so strongly? Why *should* nature seem "rational" to our brains? What are we, after all, but Johnny-come-lately primates in a universe that existed billions of years before our ancestors climbed down from the trees in Africa? Nature isn't rational or irrational, it just is. And sometime in the next decade I suspect that the quantum eraser is going to rub our nose in that fact one more time.

What Time Is It?

THE PROBLEM of defining time has been with us for a long time. Five thousand years ago, Stone Age tribes in southern England built Stonehenge, a giant astronomical observatory to indicate the beginning of the new year. Today atomic clocks keep time to better than a trillionth of a second.

From the scientist's point of view, the job of keeping time comes down to a very simple task: finding a regularly repeating phenomenon in nature and using it to make a clock. The first such phenomena were the motions of the earth — its orbit around the sun (which defines the year) and its rotation on its axis (which defines the day). Later the swinging of a pendulum took the place of the cumbersome earth, and clocks in the age of railroads kept time to an accuracy of a fraction of a second a year. More recently, the vibrations of a speck of quartz have allowed even greater accuracy.

The progression of standards from the turning earth to vibrating quartz illustrates an important truth about time: events that seem repetitive and regular according to the old standard are often seen to be jerky and irregular when examined in the fine detail provided by a newer one. The turning of the earth is actually quite erratic. When the wind blows to the west, the earth recoils to the east — an effect that can change the length of day by tens of thousandths of a second. Tides and earthquakes also can affect the length of day.

One way to designate the accuracy of a time standard is to talk about the decimal place at which it can no longer be trusted. A watch that loses one second a year, for example, would be unreliable in the seventh decimal place. By this way of reckoning, the best pendulum clocks can be trusted to six decimal places, the best quartz clocks to about nine. Neither of these is anywhere near accurate enough for modern technology.

Enter the atomic clock. In the late 1940s, physicist Norman Ramsey (who later received the Nobel Prize) devised a way of measuring very accurately the rotation of electrons in atoms. The result of this advance, the atomic clock, is the time standard used throughout the world today. In these clocks, beams of atoms wend their way through sequences of magnets and cavities in a way that allows extremely precise measurements of frequencies involved in movements of the electrons. Since 1967 the second has been defined as the time required for 9,192,631,770 repetitions of a particular cycle in a cesium atom. The advantage of using an electron as a time standard is obvious. Tides do not rise and fall and winds do not blow on electrons. Every electron in every cesium atom in the universe behaves in exactly the same way, so the cesium standard is both universal and reproducible. Atomic clocks are accurate to thirteen decimal places.

The atomic clocks that set the time in the United States are maintained at the Naval Observatory in Washington, D.C., with secondary clocks at branches of the National Institutes of Standards and Technology in Maryland and Colorado. These clocks are part of a network of atomic clocks at bureaus of standards around the world. Together these clocks keep track of the erratic rotation of the earth. Periodically they "vote" to decide whether the earth's rotation has slowed down enough to insert a "leap second" into the normal time standard. This is done every year or so, with the most recent "leap second" inserted on June 30, 1994, at 8 P.M. Eastern Standard Time.

You would think that thirteen decimal places would be accurate enough to keep anyone happy, but in fact, several efforts are under way to do better. One of these grows out of the new technology that allows us to trap single atoms (or small groups of atoms) and isolate them from the environment for a long period of time (the record is several months). Once a small group of atoms has been so isolated, lasers can slow the atoms down so that their motion has a lesser effect on the light they emit and absorb. Already such devices have, for short periods, reached accuracies of thirteen decimal places, and in principle they should be able to reach fifteen.

It has also been suggested that scientists could use radio waves emitted by the dead stars called pulsars to establish a time standard. Pulsars rotate at a rapid rate (up to thousands of times per second) and emit radio beams along their axes of rotation. The suggestion is to use the repeating signals from several stable pulsars as a time standard. Proponents claim a potential accuracy of fifteen decimal places, though little practical work has been done.

The most important use of precision timekeeping in today's technology is in the Global Positioning System, which consists of a suite of satellites equipped with atomic clocks. By looking at signals sent from at least four satellites and knowing exactly when those signals were sent, an observer can locate his or her position on the earth to within a few feet. This ability has already been used to land aircraft without the assistance of a pilot, and may someday be used to steer cars without drivers around on freeways.

How Much Is a Kilogram?

WHEN YOU BUY hamburger in a supermarket, you aren't likely to worry that the weight written on your package is incorrect. This is because there is a system stretching from your neighborhood store to scientific laboratories around the world devoted to making sure that scales are correctly calibrated. Maintaining accurate standards of measurement has always been a traditional responsibility of governments, and today it is a major scientific undertaking. But old-fashioned or modern, the basic idea is the same — the government sets the standard for weight or length or whatever, to which everyone within that government's jurisdiction must adhere.

The oldest such standard we know of is the Babylonian *mina*, a unit of weight equal to about a pound and a half. The standards were kept in the form of carved ducks (five mina) and swans (ten mina), and were presumably used in balances to weigh merchandise. In the Magna Carta, King John agreed that "there shall be standard measures for wine, corn, and ale throughout the kingdom." The marshal of the great medieval fairs at Champagne kept an iron rod and required that all bolts of cloth sold at the fair be as wide as the rod. For most of recorded history each country has kept various different standards for different purposes. In America, for example, we measure land in acres, grain production in bushels, and height in feet and inches. According to the *Handbook of Chemistry and Physics,* there are no fewer than eighteen different kinds of units called the barrel, for measuring everything from liquor to petroleum. There is even a barrel used exclusively to measure cranberries!

It was, I suppose, to get away from these sorts of confusions that the nations of the industrialized world signed the Treaty of the Meter in 1875. According to this treaty, "the" kilogram and "the" meter were to be kept at the International Bureau of Weights and Measures near Paris, and secondary standards were to be maintained in other national capitals. In the United States, they were kept at the Bureau of Standards (now the National Institutes of Standards and Technology, or NIST) in Washington, D.C. The meter was the distance between two marks on a length of platinum-iridium alloy, the kilogram the mass of a specific cylinder of the same stuff.

But since the setting of these simple, intuitive standards, advances of technology have made them obsolete. It's all very well for "the" meter to reside in a vault in Paris, but it would be much more convenient if everybody could have access to a uniform standard. Thus the trend has been away from the kind of centralized standard-keeping codified in the Treaty of the Meter and toward standards based on the one truly universal thing we know about — the properties of atoms. The development of the atomic clock is one example of such a move, the new standards for the meter another. In 1960 the platinum-iridium bar was discarded and the meter redefined as 1,650,763.73 wavelengths of a particular color of light emitted by a krypton atom. Since every krypton atom in the world is the same, this redefinition meant that every laboratory in the world could maintain its own standard meter. In 1983, following further development of the atomic clock, the meter was redefined as the distance light travels through the vacuum in 1/299,792,458 second. Again, this standard can be maintained in any laboratory.

But the kilogram hasn't changed. It's still that same cylinder sitting inside three protective bell jars on a quartz slab inside a vault in Paris. Even in such an environment, however, atoms of other substances stick to the cylinder's surface. Until 1994 it was cleaned periodically by an old technician using a chamois cloth. (I remember listening to an absolutely fascinating argument at a NIST lunch over whether or not removing atoms by washing was worse than letting gases accumulate on the surface.) When the United States wants to check whether its version of the kilogram still matches the standard in Paris, the American kilogram has to be carried overseas for tests. The last time this was done, in 1984, two scientists went with it — one to carry it, the other to catch it if it fell.

This is no way to run a high-tech society, and there is an enormous push to develop an atomic mass standard and put "the" kilogram into a museum. One technology that may allow us to do this is the new technique of isolating single atoms in a complex "trap" made of electrical and magnetic forces so that they can be studied for months at a time. These single atoms stay in the traps so long that they acquire names (the first, a barium atom trapped in Munich in the 1980s, was called Astrid). It is not too difficult to determine the mass of individual atoms to high accuracy; the problem is counting the number of atoms in a sample big enough to serve as a mass standard.

The cylinder that now constitutes "the" kilogram contains approximately 10,000,000,000,000,000,000,000,000 atoms, so even if we knew how much each one weighed to incredible accuracy, we'd have a real problem knowing how many to add. At the moment, at least five different techniques are being developed to give the kilogram an atomic definition, and I don't imagine it will be long before one of them succeeds. When this happens, the kilogram will join the meter in its museum.

So What About Time Travel?

THERE IS A long tradition of time travel in science fiction, going back to H. G. Wells's *The Time Machine* in 1895. As a kid, I was fascinated by these stories. Let me tell you about a fantasy experiment I thought about back then involving a machine that could send something (a marble, say) backward in time. I imagined a desk with two buttons and a place where the marble would appear after being sent back in time. One button would turn the time machine on and send the marble back fifteen minutes; the other would blow up both the machine and the marble. I tried to imagine what it would be like to sit at that desk and see the marble from the future pop into existence. It would be like a game of "gotcha" with the universe. If the marble didn't show up from the future, you could press the button to send it, and if it did show up you could blow up the machine and not send it. Either way, it would be a clear victory of human free will over the impersonal laws of the universe.

You run into all sorts of paradoxes when you talk about time travel. Perhaps the most famous is the "grandfather paradox," which involves someone who goes back in time and prevents the marriage of his grandfather. If he did this, of course, he couldn't have been born, in which case he couldn't have gone back in time, in which case his grandfather would have married, in which case he *would* have been born and . . . you get the idea. It was, perhaps, thinking about such paradoxes that led Stephen Hawking to propose something called the "Chronology Protection Conjecture," which says, in essence, that there will always be a reason why time travel cannot take place. As a scientific idea, time travel is a little like parallel universes. It occasionally surfaces as an unexpected consequence of some theory and is announced with great fanfare, but (so far, at least) when people look into it more closely they find a factor that hasn't been considered that will keep it from happening.

I was fortunate, I suppose, because I got involved in one of these episodes early in my career and thereby acquired a kind of immunity. In the early 1970s, physicists were talking about a hypothetical particle, called the tachyon, that could travel faster than light. It turns out that if such particles exist (and so far no one has any evidence that they do), then it would be possible to play some games with tachyons moving near the speed of light (you really don't want to know all the details) that would have the effect of sending messages into the past. This would lead to the grandfather paradox as easily as sending a person back in time — after all, the message "My God, don't marry that woman!" would start the paradox in motion as well as anything else. In this case, a student and I were able to show that the problem of distinguishing between tachyons sent on purpose and background cosmic ray tachyons would actually prevent any meaningful message from being sent by this scheme.

More recently, conjectures about time travel have concerned the massive objects called black holes, which have the effect of warping the fabric of space in their vicinity. General relativity tells us that space and time are linked (hence the term "space-time"), so distorting space also distorts time. In some cases, it is possible to find a path along which a spaceship could travel that would, in effect, bring it back to its starting point before it left.

One popular recent version of this scheme involved something called a "wormhole." Familiar to *Star Trek* aficionados, a wormhole is a hypothetical connection between two black holes — think of it as a tunnel through another dimension. The scheme involved bending the wormhole around so that the two ends were near each other, then accelerating one black hole to near the speed of light. This would have the effect of distorting time in the region and would allow a spaceship to travel "downwhen" (to use a wonderful term coined by Isaac Asimov) into its own past.

Unfortunately, when people began looking into this idea in more detail, it turned out that the situation wasn't so simple (although accelerating black holes to the speed of light isn't something most people would call "simple"). It turns out that the massive energy necessary to distort space around an accelerating black hole would disrupt the vacuum in the region and, in effect, destroy the wormhole. In other words, the "tunnel" from one black hole to another would be destroyed as soon as it was built.

Can I guarantee that every time-travel scheme will meet a similar fate? Of course not — you can never tell what future scientific theories will look like. I harbor a fond hope, for example, that human beings will find some way to circumvent the speed-of-light barrier and break out into the galaxy. I harbor no such hopes for time travel, however, despite the hours of pleasure I have derived from reading fiction devoted to it. Perhaps the most convincing argument supporting this point of view comes from Stephen Hawking. If time travel is possible, he argues, then surely a future civilization will figure out how to accomplish it.

So, he asks, "Where are all the tourists from the future?"

Gravitational Waves: A New Window on the Universe?

EINSTEIN'S general theory of relativity remains our best explanation of the phenomenon of gravity, but it has accumulated a remarkably small number of experimental verifications. This is because Einstein's relativity differs significantly from the plain old garden-variety gravity of Isaac Newton in only a few unusual situations (the degree to which light bends as it comes around the sun, for example). This means that scientists are always on the lookout for new ways to test the theory.

One prediction of the theory that could, in principle, be tested in a laboratory on earth has to do with something called gravitational waves. Whenever a mass moves, the theory states, a series of waves — small fluctuations in the force of gravity — move outward from that mass, much as waves on a pond move away from the disturbance you create by throwing in a rock. The gravitational wave created when you wave your hand through the air is much too small to be measured, but events in which large masses move quickly, as happens in the collapse of a supernova, for example, would create waves that we should be able to detect.

Here's an example to help you visualize this particular quest: imagine a long, hollow, flexible plastic tube lying on the surface of a pond. If you threw a rock into the pond, that tube would undulate as the resulting waves passed by. By watching the tube, you could detect the presence of the waves *even if you couldn't see the water*.

In much the same way, changes in a tube would signal the passing of a gravitational wave. Instead of undulating, however, the tube would register the presence of a gravitational wave as changes in its cross-sectional shape. If you looked at the tube end on, you would see its circular cross section change to an ellipse with the long axis vertical, back to a circle again, then to an ellipse with its long axis horizontal, then back to a circle, and so on.

The first (unsuccessful) attempts to detect gravitational waves in the 1970s and 1980s used large metal bars in the role of our plastic tube. The bars were heavily instrumented to monitor changes in shape, but the experiments ultimately failed because the effects to be detected were so small. The expected deviation from circular cross section of the cylinder, for example, was less than the distance across a single atom!

In experiments this sensitive, vibrations from the environment play a crucial role. Doors slamming, cars passing by outside, even the wind blowing against a building can start a metal bar vibrating, and these vibrations could easily masquerade as the effects of gravitational waves. To counter such "noise," scientists expend enormous effort to isolate their apparatus from the environment — typically, the apparatus is placed on a thick rubber shock absorber lying on a marble slab placed (if possible) on bedrock and not connected to the building in any way. Also, two experiments are run simultaneously in widely different locations, in the hope that noise at one location will not show up at the other.

In 1994 ground was broken in Hanford, Washington, for the next generation of gravitational-wave experiments. The detection scheme in this $250 million project is more sophisticated than a simple metal cylinder. The working part of the detector is a pair of 2.5-mile-long metal pipes at right angles to each other. Inside each pipe is a high vacuum, and at the end of each, in a chamber isolated from vibrations, is a test mass. Think of these two test masses as isolated chunks of a metal "cylinder" 2.5 miles long and 2.5 miles in diameter. The changes in cross section described above will, in this apparatus, manifest themselves as small changes in the relative positions of the two masses with respect to each other.

When the system is working, a beam of light from a laser will pass through a partially silvered mirror so that half of the beam travels down each of the two arms. The two beams reflect off the test masses, come back down the pipe, and are recombined. By detecting small shifts in the relative positions of the crests of the light waves, physicists will be able to monitor small changes in the positions of the test masses. A device like this is called an interferometer (because the two light beams "interfere" with each other on their return), and the entire project is called the Laser Interferometer Gravitational Observatory, or LIGO. Eventually, a second apparatus will be built in Louisiana to provide the protection against noise discussed above.

With these observatories, we will open an entirely new window on the universe. We will be able to "see" gravitational waves emitted not only by supernovae but by star systems in which the component stars revolve around each other, vibrating black holes, material falling into neutron stars and black holes, and an entire menagerie of strange stellar beasts. In the past, every time we've opened a new window on the universe — with satellite observatories, for example — we've found something new and wonderful. I expect LIGO to produce the same kind of excitement.

How Low Can a Temperature Be?

ONE OF THE great insights of nineteenth-century physicists was that temperature is related to the motion of atoms and molecules. The faster the atoms and molecules move in a material, the hotter it is. This notion leads quite naturally to the idea of a lowest possible temperature, called absolute zero, corresponding to a situation in which the atoms simply stop moving. (In quantum mechanics, this definition is modified slightly; instead of zero velocity, atoms at absolute zero are at their lowest possible energy.) Absolute zero is about −273 degrees Celsius or −456 degrees Fahrenheit, and we measure low temperatures by how close they come to this number.

There are many ways to produce low temperatures. For example, you can evaporate a liquid to draw heat away from something (this is what happens when human beings sweat), or you can expand a gas suddenly (this is why spray cans feel cold after you have been spraying for a while). Such standard mechanical techniques have been used to lower temperatures to within a few degrees of absolute zero. At that point the atoms are moving very sluggishly, but it is hard to slow them down any more.

A big breakthrough in the march to low temperatures occurred in the 1980s when scientists acquired the ability to suspend small groups of atoms in magnetic traps and manipulate them with lasers. This technique was often referred to colloquially as "atomic molasses" and involved some fancy footwork with lasers. Here is how it works: a group of atoms is trapped in magnetic fields. The atoms have a low temperature, but they are still moving a little. Laser light is shone on the atoms from many different directions, flooding them with photons. Atoms moving toward the photons will see them as blue-shifted (and therefore having higher energy), while atoms moving away from the photons will see them as red-shifted (having lower energy). The trick is to adjust the energy of the photons so that atoms moving away from them quickly will be able to absorb the light and the other atoms won't. In effect, the laser frequencies are adjusted so that only atoms moving with relatively high velocity will absorb photons.

Once the atoms have absorbed the photons, their energy is distributed around inside the atom and, eventually, radiated away. However, it is radiated at the atom's normal absorption frequency. This means that by using lasers, we have tricked the atoms into absorbing photons at low energy and radiating them at a higher energy. This energy deficit has to come from somewhere, and the only source is the energy of motion of the atom itself. Consequently, an atom that has gone through this little minuet with laser photons will wind up moving more slowly. No matter which way an atom moves, it will be overtaken by some photons, absorb them, and slow down. From the atom's point of view, it's like trying to move through a sea of molasses.

By progressively tuning the lasers, it is possible to slow the atoms down until you have a collection whose temperature is a few millionths of a degree above absolute zero. These were the lowest temperatures ever attained until quite recently. (I should mention that lower "temperatures" can be obtained in some special systems, but these involve a rather arcane definition of temperature.)

In 1995, scientists at the National Institutes of Standards and Technology (NIST) found a way to get to still lower temperatures. They started out with a collection of trapped cesium atoms that had been slowed down as described above. They then adjusted their lasers so that the light waves formed a series of troughs. Think of the atoms as a collection of marbles rolling on a table and the laser light as creating a series of troughs and peaks in the tabletop. The cesium atoms are slowed down until they fall into the troughs. Once the atoms are trapped, the lasers are adjusted so that the troughs gradually begin to widen and smooth out. When this happens, the atoms start to move away from the troughs — in essence, the collection of atoms expands. And just as a spray can gets cooler as the gas in it expands, the collection of atoms cools down as well. The extra energy is radiated away, and you get temperatures in the range of hundreds of *billionths* of a degree above absolute zero. The original experiments reported temperatures in the neighborhood of 700 billionths of a degree above absolute zero, but this number has already dropped to 30 billionths of a degree and will fall farther.

But no matter how low a temperature we can produce, no matter how much we can slow atoms down, no matter how many clever tricks we devise, we'll never actually reach absolute zero. This follows from an obscure law of physics called the third law of thermodynamics. Like the speed of light, absolute zero appears to be a fixed limit in nature that we can approach but never actually attain.

3

Astronomy
and Cosmology

What Is Dark Matter Made Of?

 IF YOU LOOK at the sky on a clear night, you can see a few thousand stars. With a telescope you can see a lot more — distant galaxies, gas clouds, and dust between the stars. All of this stuff either gives off light or absorbs it. Indeed, if stars didn't give off light, we wouldn't know they were there.

Over the past few decades, astronomers have come to realize that there is a lot of matter in the universe that doesn't interact with light (or other kinds of radiation) at all. Called, appropriately enough, "dark matter," we know it only through its gravitational effects. It is as if we could never see the moon but had to infer its existence from the ocean tides.

The presence of dark matter was first noted in the behavior of rotating galaxies like the Milky Way. The bright spiral arms of these galaxies are surrounded by a diffuse sea of hydrogen that gives off faint radio signals. By plotting these signals, astronomers can figure out the movements of the hydrogen. In every galaxy measured, an astonishing fact emerges — like leaves on a stream, the hydrogen is moving as if it were still caught up in the general flow of the galaxy, even though it's far outside the region of the stars. The only conclusion that can be drawn from this is that every galaxy is surrounded by a huge halo of material that doesn't interact with light (or any other kind of radiation) but does exert a gravitational force. When you put in the numbers, you find that at least 90 percent of a galaxy like our own Milky Way is made up of this dark matter.

Other studies have picked up dark matter between galaxies in clusters, where the gravitational force holds the cluster together much more tightly than would be expected on the basis of the gravitational attraction between the bright parts of the galaxies. It's not an exaggeration to say that we've found dark matter everywhere we've looked for it. So the universe is over 90 percent dark matter, whose very existence was unsuspected until a few decades ago.

What Is It?

In answering this question, scientists divide into two camps, which I'll label the "baryonists" and the "exotics." The term "baryon" (from the Greek for "heavy ones") refers to stuff like the protons and neutrons that make up ordinary materials. Baryonists argue that the dark matter is ordinary stuff that we just haven't detected yet — Jupiter-sized objects out between the galaxies, for example. Exotics, on the other hand, argue that dark matter has to be something new, something never before seen by humans. They pick their candidates from a stable of hypothetical particles that various theoretical physicists have dreamed up over the years but that have not yet been seen in the laboratory.

One field of battle is theoretical. In this game, scientists consider how a particular type of dark matter would fit into our theories of the early stages of the Big Bang. Would this type of dark matter be expected to give rise to the enormous structures of clusters and superclusters of galaxies we see when we look out into the heavens? Would another type explain why all galaxies seem to be so similar in size? My sense is that this argument will not be settled, because theorists on both sides will always be able to patch up their pet ideas just enough to match the (scanty) data.

So the question of what the universe is mostly made of will be decided in the good old-fashioned way, by experiments and observations. In 1993 the first gun in this war was fired when teams of astronomers announced the detection of dark objects circling the Milky Way. They had monitored millions of stars in the Large Magellanic Cloud (a small suburban galaxy near the Milky Way), watching for a star to brighten and dim over a period of a few days. The idea is that as a massive dark object comes between us and the star, light from the star will be bent around and the object will, in effect, act as a lens, making the star look temporarily brighter. Dozens of such brightenings have been seen. The dark objects are called MACHO (Massive Compact Halo Objects) and are most likely made from baryons. In 1996, astronomers announced new results that suggest that up to 50 percent of the dark matter in the Milky Way is made of MACHOs, probably in the form of burnt-out stars.

At the same time, more direct searches for dark matter are taking place in laboratories around the world. The idea of these experiments is that if we are indeed immersed in a sea of dark matter, then the motion of the earth through it should produce a "dark-matter wind," much as driving a car creates a breeze on a still day. Using blocks of perfect silicon crystals cooled to within a fraction of a degree of absolute zero, physicists are waiting for a particle of dark matter to collide with a silicon atom, depositing enough energy to disturb the crystal. These experiments are particularly looking for the exotic particles known as WIMPs (Weakly Interacting Massive Particles). If these experiments are successful, we will learn that all of our efforts in exploring the universe so far have been focused on a tiny fraction of the matter out there and that completely new forms of matter exist whose nature and effects on us are still waiting to be discovered.

How Did Galaxies Come to Be?

YOU NEED TO KNOW two things about galaxies: (1) almost all the visible mass in the universe is concentrated in them, and (2) by rights, statement 1 should not be true. This is known as the "galaxy problem" among cosmologists. And although my statement of the problem is a little flip, the question of why the universe is organized the way it is has been around (unsolved) for a long time. To see why, you have to think about how galaxies could have formed in the context of the Big Bang.

In its early stages, the universe was much too hot and dense for atoms to exist — collisions between particles would have broken them up almost instantly. This means that the material in the early universe was a plasma, like the sun, which absorbed radiation. If one part of the plasma started to collect together in a galaxy-sized clump, it would have absorbed more of the ambient radiation and been blown apart. So long as the universe was too hot for atoms (as it was for roughly the first 500,000 years of its life), galaxies could not start to form. Once things cooled off to the point where stable atoms could exist, the universe became transparent to radiation. With the impediment of absorbed radiation removed, chunks of matter could begin to come together under the influence of the gravitational force.

But that's where the problem starts. Because the universe was expanding, chunks of matter were moving farther and farther apart even as gravity tried to pull them together. If you wait until atoms can form, matter is too widely spread and moving too fast ever to be captured in galaxies.

This problem was made worse in the 1980s, when detailed

surveys of the skies revealed that not only is most visible matter collected in galaxies, but the galaxies themselves are grouped in clusters, and these clusters form still larger groupings called superclusters. Between superclusters are empty spaces called voids. Currently, the universe is thought to resemble a mound of soapsuds — the matter in superclusters is the soap film, and the voids are the empty interiors of the bubbles. But if matter couldn't have collected into galaxies during the Big Bang, it certainly couldn't have collected into the much larger clusters and superclusters.

There are basically two approaches to the galaxy problem these days — call them theoretical and observational. The former concentrates on finding some way around the dilemma, usually through the agency of an as-yet-unconfirmed substance such as dark matter. The observational approach ignores the theoretical difficulties and concentrates on searching for distant galaxies. Since light from these galaxies has been traveling toward us for billions of years, the hope is that we will eventually see a galaxy in the act of forming. With that data to point the way, observers argue, the theorists will be able to sort things out.

The most common theoretical approaches, as suggested above, invoke dark matter (see page 83). By hypothesis, dark matter does not interact with radiation and could therefore start to come together under the influence of gravity before atoms formed. In dark-matter scenarios, once the universe became transparent, visible matter was pulled into the centers of the already formed concentrations of dark matter. The mutual gravitational attraction between bits of visible matter was irrelevant — what mattered was the pull of the dark matter. Depending on the kinds of properties you assign to the dark matter, you can indeed generate galaxies, clusters, and superclusters with these sorts of theories.

While the theorists watch their computers spin out possible universes, observational astronomers are busy trying to provide data that will define the universe we actually live in. When we look at distant objects, we are also looking back in time. But distant objects are faint, which makes this search very difficult. For a long time, the most distant objects known were quasars — pointlike light sources that send out enormous amounts of energy in radio waves. But using modern electronic techniques, astronomers have started to find normal galaxies as distant as any quasar. In 1995, in fact, astronomers began to find galaxies that seem to be in the act of forming — light must have left these so-called primeval galaxies when the universe was less than a billion years old. As these kinds of data start to come in, theorists are already having to revise their dark-matter scenarios, many of which have problems producing galaxies that quickly.

I expect this sort of interplay between theory and observation to quicken in the near future, as more data on distant galaxies accumulate. The hope, of course, is that the data will winnow out theories that don't work, and we will be able to remove statement 2 from the list above.

How Old Is the Universe?

 IN 1929 the American astronomer Edwin Hubble forever changed our view of the universe. Using the then state-of-the-art telescope on Mount Wilson, near Los Angeles, he proved that (1) stars in the universe are clumped into galaxies, and (2) the galaxies are moving away from each other. He discovered, in other words, that the universe is expanding.

From this discovery it is a short jump to the idea of the Big Bang, the notion that the universe began at a specific time in the past and therefore has a specific age. (Think about the current expansion and imagine running the film backward until everything shrinks down to a single point.) A little-known historical sidelight of Hubble's original work is that his first value for this age was a mere 2 billion years. From the early 1930s until 1956 (when, as discussed below, an inconsistency in the original measurement was cleaned up), the age of the universe was officially about half that of planet Earth! I mention this historical anomaly because today, thanks to results from the orbiting telescope named after Hubble, a very similar situation seems to be developing.

First a word about finding the age of the universe. For each galaxy out there, you have to know two things: how fast it's moving away from us and how far away it is. The actual calculations are analogous to (but slightly more complicated than) the problem of knowing a car is traveling twenty miles per hour and is sixty miles away, from which you can deduce that the trip started three hours ago.

Measuring the speed of a galaxy is easy. You look at the light coming from an atom in that galaxy and see if it has the same wavelength as light emitted by the same kind of atom in your laboratory. If the light from the galaxy has a longer wavelength (and thus is redder), the source must have shifted between the time of emission of one crest and the next. Measuring that shift and dividing by the time between pulses gives the speed of the atom and of the galaxy it is part of.

It's not finding the speed of the galaxy, then, but finding the distance to it that has been (and continues to be) the main problem with discovering the age of the universe. Looking at a distant galaxy is like looking at a light bulb in a totally darkened room — you can never tell if you're seeing something bright and far away or dim and very close.

The way astronomers have always attacked this problem has been by finding a "standard candle," something whose total output of energy is known. The idea is that if you know how much energy the object emits and measure how much you actually receive, you can calculate the distance to it. In the analogy of the darkened room, a 100-watt light bulb would be a standard candle — if it looked faint, you would know it was far away, if it looked bright it would be close.

Historically, the most important standard candle in astronomy has been a type of star called a Cepheid variable (the name derives from the constellation Cephus, where the first such star was seen). These stars go through a regular cycle of brightening and dimming that lasts several weeks or months, and the length of the cycle is related to the total energy that the star radiates into space. Measuring the cycle time of a Cepheid, then, is equivalent to reading the wattage rating of the light bulb in our example.

It was Hubble's ability to see individual Cepheid variables in nearby galaxies that made possible his discovery that the universe is expanding. It was the discovery in 1956 of two different kinds of Cepheids (which Hubble had mixed together) that later made the estimated age of the universe older than the age of the earth. Up to 1994, we didn't have techniques powerful enough to allow us to see individual stars in clusters of galaxies more than about 50 million light years away, which meant that indirect estimates had to be used for distances to galaxies from that point out. The result was that estimates of the age of the universe ranged from 7 to 20 billion years.

In 1994, two groups of astronomers (one operating from the ground with advanced electronic systems, the other using the Hubble Space Telescope) finally isolated some of these stars and determined the exact distance to a group of galaxies called the Virgo Cluster. The results of their measurement: the universe is between 8 and 12 billion years old.

There's still a lot of controversy about these numbers, so don't be surprised if other scientists come up with older ages in the next few years. If this age range stands, however, it is going to shake up a lot of our ideas about how the universe works. The most pressing problem is that astronomers who work on the life cycles of stars tell us that large numbers of stars in the universe are between 12 and 16 billion years old. My own guess is that we're in for a replay of history, that we will find it necessary to raise the age of the universe. If it turns out that that can't be done, then the whole Big Bang picture of the universe may have to be reexamined.

What fun!

Where Are the Sun's Missing Neutrinos?

 EVERY SCIENCE has a problem that can't be solved and just won't go away. For astronomers who study stars, it's the solar neutrino problem.

Neutrinos are wraithlike particles produced in nuclear reactions. They have no mass or charge, and they scarcely interact with matter at all. Millions of them are passing through your body as you read this, but they don't interact with your atoms so you never know they're there. In fact, a neutrino could travel through a sheet of lead several light years thick without disturbing anything or leaving any trace of its passage.

Their elusiveness makes neutrinos hard to detect, but it also makes them enormously useful to people who study the sun. We believe we know what nuclear reactions are going on in the core of the sun, and this means we know how many neutrinos are being made. These neutrinos pour out of the center of the sun and cross the orbit of the earth about eight minutes later. And although the probability of any given neutrino interacting with an atom on the earth is small, there are so many of them that just by chance there will be an interaction every once in a while.

These fleeting collisions, in fact, are what give rise to the solar neutrino problem. Starting in the mid-1960s, an experiment has been running in a chamber 4,800 feet underground in the Homestake Gold Mine in Lead, South Dakota. The main piece of the experiment is a tank holding more than 600 tons of cleaning fluid. A few times a day, one of the neutrinos flooding through the tank hits a chlorine atom in the cleaning fluid and converts it into argon. Using precise chemical methods, scientists gather and count the argon atoms, one by one. And since the mid-1960s, their counts have always come up short — we're only getting from one third to one half of the neutrinos we should be getting from the sun. More recently, experiments in Russia and Europe, using different materials and different techniques for counting neutrinos, have come on line, and the results have been the same: for the past quarter century, the number of neutrinos passing through the earth on their way out from the sun has been too low.

At first, people didn't worry too much about this finding — after all, the South Dakota experiment could have been wrong, and possibly some of the details in our understanding of nuclear reactions in the sun were wrong. But as time went on and the loopholes in the interpretation of the results began to be closed, astronomers became more and more concerned. After all, if we can't understand how our own sun works, despite the enormous amount of data we have on it, how can we possibly hope to understand the enormous variety of other stars in the sky? Some of the proposals began to take on a note of desperation. I recall one suggestion, for example, that there was a growing black hole in the center of the sun that had already consumed a chunk of the sun's core, where nuclear reactions take place, and would someday eat its way to the surface to produce what can only be called the ultimate environmental catastrophe.

Today, hopes for a resolution of the solar neutrino problem are riding on the properties of neutrinos themselves. We know that there are three different kinds of neutrinos, called the electron, mu, and tau neutrinos, each of which is associated with a different fundamental particle. If the neutrino's mass is exactly zero, then theory tells us that these three different types must forever stay separate. Since only the electron neutrino is produced in nuclear reactions and can interact with atoms on earth, then in the conventional scheme of things the existence of the other two neutrinos is simply irrelevant to the solar neutrino problem.

If, however, the neutrinos have a small mass (as some modern unified field theories predict), then a process called mixing or "oscillation" can take place. As the neutrinos stream out of the sun, they start to change identity. What was originally a stream of electron neutrinos becomes a stream that contains all three types. Think of a stream of cars entering a freeway. Initially, all the cars are in the right-hand lane, but as time goes by, some move into the other lanes and only a fraction remain on the right. Since only electron neutrinos can trigger detectors on earth, and since fewer of that sort of neutrino will reach us because of the oscillations, the solar neutrino problem would be resolved.

This notion got some support in 1995, when scientists at Los Alamos National Laboratory announced the results of an experiment in which beams of mu neutrinos were shot into a twenty-eight-foot tank filled with baby oil. They saw a few events initiated by an electron neutrino, which argues that neutrinos do indeed oscillate (in this case from mu to electron types) — assuming, of course, that the experiment is right. Plans are afoot to improve on this experiment; one proposal calls for directing a neutrino beam through the ground from the Fermi National Accelerator Laboratory near Chicago to a detector in northern Minnesota and see what happens to the beam on the way. The solar neutrino soap opera has been going on for so long, though, that I'm losing hope for a simple solution.

How Will the Universe End?

EVER SINCE Edwin Hubble showed that the universe is expanding, we have known that it began at a specific time in the past (although we may still argue about precisely when that event occurred). Today cosmologists routinely discuss events that took place in the universe's first microsecond of existence. We seem to have a pretty good sense of the main features of the beginning of the universe.

We can't say the same about the future of the universe, however. The basic question is easy to frame: will the expansion going on now continue forever, or will it someday reverse itself? If the universe is destined to expand forever, it is said to be *open*. If the expansion will someday reverse, so that the galaxies fall together in what astronomers (only half-jokingly) call the "Big Crunch," the universe is said to be *closed*. The border between these two cases, in which the expansion slows down and comes to a stop over an infinite amount of time, is called a *flat* universe. Open, closed, or flat — those are the options among which we must choose if we want to understand the fate of the universe.

A simple analogy helps in understanding this question. If you stand on the surface of the earth and throw a baseball upward, it will eventually fall back down. This is because the entire mass of the planet underneath your feet is exerting a gravitational pull on that baseball, causing it to slow down, stop, and come back to the ground.

If you threw a baseball upward on a small asteroid, however, you could easily throw it out into space. Throwing it no harder than you had on the earth, you would see it escape from the asteroid, never to return. It's not that the asteroid wouldn't exert a gravitation force on the baseball, for it would. It's just that the asteroid doesn't have enough mass to reverse the baseball's direction. The baseball might slow down a bit, but it wouldn't stop and fall back.

If you think of a distant galaxy as analogous to the baseball, then the question of whether the universe is open or closed comes down to whether there is enough mass in the rest of the universe to make that galaxy reverse its motion. The question of the fate of the universe, then, is an observational one. If there is enough mass around to exert a gravitational force capable of pulling that galaxy back in, the universe is closed. If there's not enough mass, the universe is open. It's as simple as that.

Well, it's actually not simple in practice, because finding out how much mass there is in the universe can be pretty tricky. You could begin by counting up the mass of all the matter that you can see — stars, nebulae, even dust clouds that we "see" because they absorb light. This so-called luminous matter is only about 0.1 percent of the amount of mass needed to "close the universe" (to use the astronomer's jargon). Add in all the dark matter we know about because of the way its gravitational pull affects the motions of galaxies and galactic clusters and the number goes up to about 30 percent.

The big question is whether there is more mass. If there isn't, then the universe is open and the current expansion will go on forever. Over tens of billions of years the stars will stop shining as the primordial hydrogen from which they derive their energy is used up. Theoretical models suggest that over unimaginably long times, all the matter in the universe will either decay into elementary particles or be converted into black holes. The black holes, via a phenomenon known as Hawking radiation, will then be converted into electrons and positrons traveling through empty space.

The currently fashionable theories of the origin of the universe ("inflationary universes") predict that the universe will ultimately turn out to be flat. That is, there will be just the critical amount of matter for the universe to be balanced between eternal expansion and eventual contraction. You often see the statement that the universe "must" be flat and hear references to the "missing mass." This is an unfortunate turn of phrase, because the mass would be "missing" only if we knew for certain that it was supposed to be there.

There are plenty of places in the universe where more dark matter could be hiding. Many theories say that the concentration of dark matter is something like a mountain range. In most places the concentration is low, but there are occasional peaks. On those peaks luminous matter (such as galaxies) collects like snow on the Rockies, but the majority of dark matter is actually spread thinly between the peaks, where, according to these theories, we haven't been able (as yet) to detect its presence. If that is where most of the dark matter is, then the universe could well be flat. In that case its future wouldn't be all that different from the future of the open universe — a gradual running down of creation.

Not a whole lot to look forward to, is there?

How Many Other Stars Have Planets?

THERE ARE lots of reasons why you might want an answer to this question. From a purely scientific point of view, finding other planetary systems gives us insight into our own — is it ordinary or unusual, common or unique? And then there is the question of life. Since the only life we know about developed on our planet, it's pretty clear that if we ever expect to find a real-life version of the bar scene from *Star Wars* we'll have to find more planets somewhere.

The central problem is that on the astronomical scale, planets just aren't very big or very important compared to stars. To someone looking at the solar system from even the nearest star, none of the planets would be visible as separate objects. To understand why, think about this problem: the brightest light in the United States is the beacon on top of the Empire State Building in New York. Suppose someone placed a birthday cake with one candle right at the edge of the beacon's lens. Seeing Jupiter from the nearest star would be like trying to see that candle from Boston!

What this means is that those searching for planetary systems have to look for something other than images of individual planets. The most successful strategy is to look for wobbles in the movement of a star caused by the gravitational pull of the unseen planets.

Detecting the effects of planets on the motion of a star is extremely difficult. Think about the influence of Jupiter on the sun as an example. We are used to thinking of Jupiter as going around the sun, but in fact the two actually move around a point in between, like partners in a waltz. Thus, watching the sun should be like watching an out-of-balance washing machine on spin cycle — there should be a wobble. But the point about which the sun and Jupiter rotate (called the center of mass) is actually only 1,000 miles from the sun's center — well inside the body of the sun itself. This means that the sun's wobble would be very small and very hard to detect. If detected, however, such a wobble does indicate that there is a planet, for nothing else could produce it.

Since the 1930s, claims for discoveries of planet-induced wobbles have cropped up occasionally. The outcomes have been depressing — for a while everybody was excited, then the second round of measurements or interpretations would come in and the claim was either retracted or discounted. In fact, it wasn't until 1994 that generally accepted evidence for another planetary system came in.

It came from measurements of radio waves emitted by a pulsar — the small, dense, rapidly rotating remnant left after a big star explodes in a supernova. This particular pulsar (called PSR B1257+12) rotates about 160 times a second, beaming a pulse of radio waves at the earth on each turn. If there are planets affecting the path of the pulsar, the time between pulses will appear to decrease when the star moves toward us, increase as it moves away. Scientists monitoring PSR B1257+12 have detected irregularities they say are caused by planets. In fact, they claim that the pulsar is accompanied by three planets — two that are about three times the size of the earth, one about the size of the moon.

But you have to remember that a pulsar is what's left after a supernova. An explosion like that would surely destroy any planet that had originally formed around the star. In fact, theorists suggest that the pulsar's planets formed from debris left over from the supernova and therefore couldn't possibly have life on them. In other words, *the first confirmed planetary system we found was around the wrong star!*

Then in 1995 a team of Swiss astronomers announced the discovery of the first planet circling an ordinary star. This was a planet in orbit around 51 Pegasi (the fifty-first brightest star in the constellation Pegasus). Intermediate in size between Jupiter and Saturn, this planet orbits so close to its star that it almost touches the star's outer atmosphere; it would be well inside the orbit of Mercury if it were in our own system.

At a meeting in San Antonio in 1996, two American astronomers made an announcement that showed, as most astronomers expected, that planets are actually quite common in the Milky Way. Monitoring stellar wobbles at the Lick Observatory near San Francisco, they found two more stars with planets. The star 70 Virginis had a huge planet — eight times the size of Jupiter — in an orbit that would be just outside Mercury's in our solar system. The other planet, circling the star 47 Ursae Majoris, was only about three times the size of Jupiter and was in an orbit that would be outside that of Mars. For a while this planet was nicknamed "Goldilocks" because it wasn't too hot or too cold, but just the right temperature to have liquid water in its atmosphere.

With these discoveries, the race was on to find more planets. As this book goes to press in the summer of 1996, one more discovery has been announced and astronomers are starting to talk about the "Planet of the Month Club." More important, our current theories about solar systems tell us that giant planets like Jupiter should form only in the outer reaches and that planets close to the star should be small and rocky like Earth. Unless astronomers come up with a theory that the newly found planets formed farther out and then moved closer to their stars, we may have to rethink the process of planet formation.

Is Anybody Out There?

 I CAN ALWAYS tell when something new is happening in the search for extraterrestrial intelligence (SETI) because I start getting calls from reporters. Years ago, in conjunction with my friend and colleague Robert Rood, I wrote a book that argued against the predominant view that the galaxy is full of intelligent life. I am obviously in a lot of Rolodexes as the guy to call for responsible dissent, but I'm afraid a lot of the reporters go away disappointed. The reason: although I believe we are probably the only advanced technological civilization in the galaxy, I believe that SETI should be pursued vigorously. It is one of the few scientific enterprises I can think of whose results would be just as interesting if they came out negative as if they came out positive.

Human beings tend to populate the skies with alien beings. For a time in the nineteenth century, for example, we believed that people lived in the sun — the thought was that beneath its fiery exterior, the sun was pretty much like the Earth. There was even a "SETI" program designed to look down through sunspots and see them. The science fiction of the 1930s and 1940s was full of bug-eyed monsters from other planets (who usually had a strange affinity for scantily clad starlets). Since those naive early times, our thinking about SETI has gone through two distinct stages.

The first stage began in 1959, when a group of scientists gathered at the Green Bank Radio Astronomy Observatory in West Virginia to discuss the possibility of detecting the presence of extraterrestrials. This conference produced the famous

Drake equation (named after astronomer Frank Drake), which says that the number of extraterrestrial civilizations in the universe trying to communicate with us *right now* is

$$N = R \times P \times E \times L \times I \times T$$

where R is the number of stars in the galaxy, P the probability that a star has planets, E the number of planets per star capable of supporting life, L the probability that life will actually develop, I the probability that intelligence will develop (intelligence being defined as the ability to make radio telescopes and send signals), and T the length of time that the signals will be sent. By guessing at the values of these variables, the scientists estimated that there were literally millions of life forms in the galaxy. This idea, of course, permeated popular culture and is now accepted as absolute truth.

The second stage of thinking about SETI developed as we began to look for all of those millions of ETs. By the 1950s the idea of life on the sun had been dropped, of course, but it was felt that we would surely find life on Mars and Venus, and perhaps on some of the larger moons as well. As the exploration of the solar system unfolded over the last quarter century, however, these hopes were dashed. Venus turned out to be a torrid hell, Mars a cold, waterless desert. Despite the optimism of the 1950s, there is today no evidence for life anywhere in the solar system, although there may once have been primitive life on Mars.

Furthermore, as we have come to understand the dynamics of systems such as planetary atmospheres, we have come to realize that planets like Earth — planets that have liquid water on their surface for the billions of years it takes advanced life forms to develop — are probably quite rare. The key concept here is the "continuously habitable zone" (CHZ), a band around a star in which a planet like the earth can exist. The

CHZ around the sun extends from orbits 1 percent smaller than that of Earth to orbits 5 percent larger. Had Earth been farther out, it would have frozen over long ago; if it were closer, it would be like Venus. Stars smaller than the sun have no CHZ at all, and those larger don't live long enough for life to develop.

Even planetary size plays a role in the development of life. Planets bigger than Earth have too many volcanoes and become like Venus; planets that are smaller lose their atmospheres and become like Mars. Also, calculations indicate that planets without large moons have axes of rotation that flip around chaotically — a process that would surely wipe out life should it occur on Earth. So to find intelligent life, we need to find the right-sized planet, with a large moon, circling at just the right distance from a right-sized star. It may well be that Earth is the only planet in the galaxy that meets those conditions!

But even if you think the odds of success are small, SETI has to be tried. The main technique used in the past (and in the foreseeable future) is to look for radio signals sent either intentionally or unintentionally by other civilizations. A number of such searches have been made (all unsuccessful so far), but no large-scale, systematic survey has been done. Congress has been very reluctant to fund large-scale projects in this area. To its credit, the NASA bureaucracy has made a real attempt to keep the money flowing — even to the extent of giving the program a more opaque name. But Congress realized that HRMES (High Resolution Microwave Survey) was just SETI in another guise, and axed it in 1993. Now, phoenixlike, SETI is rising from the ashes, supported by private money. The search will go on.

How Variable Is the Sun, and Does It Matter?

 ALMOST ALL analyses of the earth's climate proceed on the assumption that the only systems of interest are internal to the earth itself — greenhouse gases, Antarctic ice sheets, ocean currents, and so on. These analyses ignore the single most important driving force of the earth's climate — the sun. Until quite recently, almost no one asked whether the energy we receive from the sun varied substantially over time.

We have had direct measurements of the "solar constant" — the amount of solar energy that arrives at the top of the earth's atmosphere — since the late 1970s, when satellites started to record this information. We knew that the number of sunspots on the sun goes through an eleven-year cycle (you may be aware of this, because during the peak of the sunspot cycle, particles from the sun sometimes disrupt radio transmissions on the earth). Scientists expected that the brightness of the sun would follow the same cycle, being lowest when there were the most sunspots and highest where there were few. The total expected change in brightness wasn't large (perhaps 0.1 percent), but even such a small change could well play a role in the earth's climate.

To everyone's surprise, it turned out that the sun is brightest when there are the most dark spots on its surface. There are, in fact, two kinds of spots on the solar surface — the familiar sunspots, which darken the sun's disk, and hot spots called faculae, which brighten it. The number of spots of both kinds increases and decreases at the same time over the eleven-year cycle, but the bright spots "win," so the sun is actually brightest when the greatest proportion of its surface is covered with dark spots. The important question for climatologists is whether the energy we receive from the sun varies over time scales of decades or centuries in any way other than the sunspot cycle. If it does, then solar warming or cooling will surely have an effect on the climate, and future theories of global warming and related phenomena will have to take that into account.

There are two ways of getting at this question — one astronomical, one geological. Astronomers don't have direct data on the sun's brightness over time periods of centuries, but they can monitor a number of stars like the sun for shorter periods and try to put together a picture of stellar behavior that will allow us to make statements about the past and future behavior of our own star. This technique is like doping out the life cycle of trees from photographs of a forest. These sorts of astronomical studies indicate that the variation in the range of 0.1 percent in the sun is about what stars like the sun usually produce.

Geological measures of solar activity depend on the fact that the earth's atmosphere is always being bombarded by cosmic rays and that the planet's magnetic field deflects them, serving as a partial shield. During times of maximum sunspot activity, material flowing out from the sun distorts the earth's magnetic field, and the influx of cosmic rays increases. These fast-moving particles, when they collide with atoms in the atmosphere, produce a number of distinctive isotopes, including carbon-14 and beryllium-10 (beryllium is a light metallic element similar to magnesium). The beryllium-10 comes out of the atmosphere in snow and is incorporated into ice packs. Scientists can drill into the ice, measure the amount of beryllium deposited long ago, and reconstruct the sun's activity in the past. The results are striking: over the last century and a half, the sun has gotten warmer, which could account for as much as 0.25 degrees Celsius of the earth's warming. In other words, changes in the sun could have as big an effect as human-generated carbon dioxide has had!

In general, scientists are reluctant to accept this sort of result. Not only would it mean a rather embarrassing retraction of public pronouncements about global warming, but there is a long historical record of incorrect theories about the sun's effect on the earth's climate. Since the existence of sunspots was first established in 1851, the sunspot cycle has been invoked to account for everything from the price of grain in England to the severity of the monsoon in India. To show how easy it was to make these kinds of correlations, I once did a "study" in which I showed that the length of women's skirts in the United States during the twentieth century followed the sunspot cycle almost exactly (periods corresponding to the flappers of the twenties and to the miniskirts of the sixties were two extreme data points in this analysis). No wonder scientists, who are at heart conservative creatures, view any attempts to explain weather in terms of the sun with skepticism!

Over the next few years I expect climate modelers will start to include changes in the sun in their calculations. We will also get better at reconstructing past solar brightness. I wouldn't be surprised if this sort of work completely changed our outlook on global warming. After all, Congress may be able to pass laws about carbon dioxide emissions, but it's a little harder to regulate the sun!

What Are Quasars?

TO BE perfectly frank, I didn't expect to be asking this question when I began writing this book in 1994. Quasars are strange things, as we'll see, but a beautiful and elegant theory explained what they were and tied them in to other phenomena in the universe. Unfortunately, in 1995 new data from the Hubble Space Telescope challenged this understanding and reopened the whole question of the identity of quasars. Another beautiful theory destroyed by an ugly fact!

The word "quasar" is a contraction of "quasi-stellar object." First seen in 1963, quasars are so far away that they looked like points of light, even when viewed through the best telescopes then available. Quasars were then the most distant objects we could see. The fact that we can detect them despite their distance from us means that they must be pouring enormous amounts of energy into space — as much as 10,000 times the energy emitted by a galaxy like the Milky Way. To round out the picture, quasars are observed to flare up and die down over a period of days or weeks, which means that their energy source, no matter how powerful, must be contained in a volume only a few light days or light weeks across. (For reference, the Milky Way is somewhat less than 100,000 light *years* across).

Some quasars are believed to be over 14 billion light years from the earth. Just as someone approaching a city at night in an airplane will see from a distance only the brightest lights of downtown, astronomers peering into the depths of space will see only the most luminous objects. Since light from these quasars has been traveling for most of the lifetime of the universe, we are seeing quasars as they were billions of years ago, not as they are now. For a while, astronomers thought that quasars might represent some violent early phase in the evolution of galaxies — a state every galaxy passed through on its way to stodgy normalcy. Many writers (myself included) speculated that astronomers in a distant normal galaxy that once was a quasar might, in looking at the Milky Way, see a quasar in the light it emitted billions of years ago.

As telescopes got better at picking out faint objects in space, however, this evolutionary view of quasars lost favor, because astronomers were able to detect normal galaxies that, although fainter than the quasars, were equally distant. Whatever quasars are, they seem to be part of the galactic zoo, not just an evolutionary phase.

In fact, it seems there are two distinct sorts of galaxies in the universe. By far the greatest number are like our own Milky Way — homey, friendly places where stars go through their evolution and life could (in principle, at least) develop. But some galaxies — perhaps a few tens of thousands in all — are very different. They are violent places, wracked by explosions, pouring huge amounts of energy into space. Quasars are the epitome of these so-called active galaxies, but there are other kinds as well. Called radio galaxies, Seyfert galaxies, and BL Lac objects, they all display the same massive energy output and rapid variation as quasars, although none are quite as energetic.

The theory that eventually developed to explain active galaxies was beautiful and simple. The idea was that at the center of these galaxies there was a huge black hole — for a quasar, one with as much as a billion times the mass of the sun. Material falling into this black hole from the surrounding galaxy would heat up and radiate its energy into space. The flareups in brightness would be caused by largish clumps of matter falling in and causing a temporary brightening. In this scheme, different types of active galaxies would correspond to different-sized black holes. The theory explained most of the observed details of active galaxies in terms of the fact that our viewing angle for each black-hole complex was different. The black-hole-plus-in-falling-matter-plus-host-galaxy would look different if we were looking down on the system than if we were looking at it end on.

But every theory in science, no matter how beautiful and compelling, has to meet the test of observation. A crucial piece of this picture is that every quasar is supposed to be a black hole surrounded by a host galaxy to feed material into it — the brighter the quasar, presumably the larger the host galaxy. After the Hubble Space Telescope was repaired, astronomers had, for the first time, an instrument that could examine quasars in enough detail to see those host galaxies.

But it didn't. In what has to be one of the most surprising events in modern astronomy, scientists announced in 1995 that their first survey had turned up a large number of "naked" quasars — quasars in which the central bright spot didn't seem to be surrounded by anything. Of the fourteen galaxies in the initial survey, eight showed no trace of a host galaxy at all, and only three had host galaxies of the type predicted by the theory. Since that time, repeated observations have turned up faint galaxies on some (but not all) of the "naked" quasars.

So we're back where we started. The oldest, most distant things in the universe are once again among the most mysterious.

So Where Are All the Black Holes?

NO DOUBT about it, the idea of black holes is one of the weirdest ever conceived by the mind of man. Theoretical physicists trying to solve Einstein's equations of general relativity in the early twentieth century were the first to stumble across it. The theory seemed to predict the existence of objects in which matter was so compacted that nothing, not even light, could escape from their surfaces. At first, many people believed that black holes were merely figments of the theoretical imagination, but over the last few decades people have started to take the notion seriously and to ask whether they exist in nature.

There are a number of ways to picture black holes. The simplest way is as masses crammed into so tiny a space, exerting so strong a gravitational force, that if you shone a flashlight up from the surface, the light could never escape. Or you could be a little more abstract and say that the mass is so concentrated that it wraps space around it like a blanket. Any way you think about a black hole, though, you get a place cut off from the rest of the universe. Any light falling on it will never come out, which means it will look black (hence its name).

Scientists talk about three different kinds of black holes, and it may come as something of a surprise to learn that we have conclusive evidence for only one of these in nature.

Quantum Black Holes

Some theories suggest that all of space (including the space surrounding you at this moment) is dotted with tiny black holes. Much smaller even than the particles inside the atom, these black holes are supposed to provide a foamy backdrop to more familiar objects. Although a few theorists (most notably Stephen Hawking) have suggested that such things might exist, there is no evidence that they do. Indeed, I'm not sure anyone would even know how to look for them. For the moment, quantum black holes are merely a gleam in the theoretical eye.

Stellar Black Holes

This is what most people mean when they use the term "black hole." The idea is that when very large stars — thirty or more times as massive as the sun — become supernovae and then, under the crushing force of their own gravity, collapse, they pull space in around them as they go. These objects would be enormously massive but only a few miles across.

If a stellar black hole sat out in space all by itself, it would be virtually impossible to detect from the earth — what could you possibly look for? So the search concentrates on finding double star systems in which one star is visible and the other (unseen) companion is a black hole. By watching the motion of the visible partner, you can deduce the properties of the unseen companion. The best candidate so far is a star system called Nova Perseii 1992 ("new star in the constellation Perseus in 1992"). This is a star orbiting an unseen companion that is too massive to be anything but a black hole. The system occasionally emits bursts of X-rays (presumably the result of material falling into the black hole), which is how it came to the attention of astronomers. We now know of a half-dozen double star systems that might be stellar black holes.

Galactic Black Holes

Finally, there is very strong evidence of huge black holes at the very centers of galaxies. As of this writing, the largest of these weighs in at 40 million times the mass of the sun. As with stellar black holes, these objects can't be seen directly, but we can see how they affect their neighbors. What we see is clouds of gas circling around the centers of galaxies. If there are only stars at the center, the closer gases will move more slowly than the stuff farther out. If the clouds are circling around a black hole, on the other hand, then the closer you get to the center the faster they will be moving. This is a pretty clean signal for the black hole.

Scientists working with ground-based telescopes have already accumulated evidence for these large black holes in some galaxies, including our own Milky Way (whose central black hole is of modest size, perhaps less than a million solar masses). With the repair of the Hubble Space Telescope, however, the search can be extended, because the instrument can see the centers of distant galaxies in very fine detail. The question that astronomers will be asking over the next decade: does *every* galaxy have a black hole at the center?

The discovery of galactic black holes, then, will add to our detailed knowledge of the structure of galaxies, and the confirmation of stellar black holes will strengthen our understanding of the life cycle of massive stars. But the most important result of evidence proving the existence of black holes, I think, will be the affirmation that even the wildest theoretical predictions have a way of turning up as fact in this strange universe we inhabit.

Parallel Universes and All That

HAVE YOU ever wondered whether there was a universe where some parallel you was reading a parallel book, except that the pages of the book were purple and you had three eyes? If so, then you are already aware of how attractive the notion of parallel universes is for the human imagination. What you may not know is that this notion also is attractive to theoretical cosmologists. In fact, the concept of parallel universes seems to move in and out of accepted cosmological theory as rapidly as changes in the width of men's ties or the length of women's hemlines.

Some background: the science (as opposed to science fiction) associated with parallel universes comes from the kinds of theories of the origin of the universe discussed elsewhere in this book (see page 9). All of these theories share a common feature — the universe arose from the primordial vacuum because that vacuum was unstable in some way. Different theories assign different sorts of instability to the vacuum, and therefore predict universes with different properties. In general, however, they share one notion: that at the creation of the universe, the fabric of space-time became highly warped and distorted, and the energy involved in this distortion was eventually converted, by a series of high-energy interactions between elementary particles, to the mass we see around us today.

The reason that the notion of parallel universes seems to go in and out of fashion is that cosmologists never construct theories to deal solely with this phenomenon. They are always trying to adjust a theory to produce the right mass density for the universe, for example, or the right proportion of dark matter. In each case, the question of whether or not the theory also predicts parallel universes is something of an afterthought. It is, however, often the only aspect of the theory seized on by science writers. When the original theories are abandoned (again, for reasons that have nothing to do with parallel universes), the notion fades from consciousness until the next set of theories comes along. In a sense, then, whether or not we are thinking about parallel universes at a given time is an accidental consequence of the prevailing cosmological theories.

Let me give you an example of how a theory of the universe might lead into this kind of cycle. One common analogy used to describe the early universe is to imagine it as an expanding balloon. In the standard picture of the expanding universe bequeathed to us by Edwin Hubble, the surface of the balloon is smooth and unwrinkled, which in theoretical terms means that very little energy is tied up in the surface itself.

In the mid-1980s, however, some physicists exploring ideas in the field of elementary particles suggested that at the high temperatures that obtained during the first fraction of a second of the life of the universe, some versions of their theories predicted interactions among elementary particles that could change this picture. In essence, they suggested that interactions between elementary particles could produce microscopic (but very intense) wrinkles on the balloon's surface. If the surface was wrinkled enough, small regions could neck off from the main balloon to form their own little bubbles. According to standard cosmological theories, those small bubbles, given time, would start expanding and become large balloons in their own right.

Voilà — parallel universes! In this picture, every expanding universe sheds baby universes the way a dog sheds hair in the summer. Unfortunately, as time went on, the original theories of elementary-particle reactions were superseded (for reasons you really don't want to know about), and at present you don't hear much about parallel universes.

In a way, that's too bad, because I think you could ask a very interesting question: could life develop in any of these parallel universes? We know that, given the right conditions, life can develop quite rapidly on a suitable planet. But if there was too much mass in those baby universes, the universal expansion might turn into a contraction before stars could form. Or maybe the charge on the electron in those universes was too small for atoms to form, in which case there would be no atoms, no chemistry, and no life. These sorts of speculation can be (and have been) worked out in great detail. We know pretty well what the necessary conditions are for life in a universe, but to my knowledge no one has ever tried to apply this thinking to a particular parallel-universe theory to see how prevalent life (even in theory) might be.

Oh well, I suppose it really doesn't matter all that much, because parallel-universe theories have one other point in common: all of them say it is impossible for any sort of communication to take place between one universe and the next. So even if there is a parallel universe out there, we'll never know about it!

Why Is Pluto So Weird?

IT'S A GOOD THING Isaac Newton never heard of the planet Pluto. If he had, he might have had second thoughts about comparing the solar system to a majestic, orderly clock. From the moment of its discovery, Pluto has been odd man out among the planets.

All the planets circle the sun in the same plane — except Pluto, whose orbit is tilted at an angle of 17 degrees. All the planets have orbits that are roughly circular — except Pluto, whose orbit is so elliptical that between 1979 and 1999 it was actually inside the orbit of Neptune. The planets display a regularity in their composition, with the inner planets being small and rocky, the outer planets large and gaseous — except Pluto, which is a small rocky world out at the edge of the solar system.

I could go on, but you get the idea: Pluto just doesn't seem to fit. This situation naturally led to speculation that it might not be a planet at all but either a captured comet or a moon that had somehow broken free. In 1978 astronomers found that Pluto has its own moon (named Charon, after the man who, in Greek mythology, ferried the dead into the realm of the underworld). This discovery has made both of these notions harder to defend. The chance of a single object being captured into the solar system is small enough, and the chance of there being two is that much smaller.

Astronomers these days are starting to look at Pluto not as a planetary freak but rather as a normal resident of the outermost regions of the solar system. These regions have always been something of a mystery because (1) we have not had telescopes capable of seeing the small objects out there and (2) we have not had computers capable of dealing with the extremely complex and shifting gravitational forces that these objects are subject to. Over the last five years, however, the Hubble Space Telescope and new computers have changed the situation.

The emerging picture is that when the planets were forming, the outer planets, through their gravitational influence, prevented the formation of planets both inside the orbit of Jupiter (hence the asteroid belt) and outside the orbit of Neptune. What lay outside Neptune in the beginning was a more or less uniform disk of objects varying in size from a few miles to a few hundred miles across. Over time, various influences have depleted the population of these objects. For example, the objects sometimes collide with each other, kicking both of them out of the disk. Passing stars provide perturbation, as do the gravitational influences of the gas giants. Calculations indicate that there should be a gradual depletion of objects on the inner side of the disk (astronomers call this "gravitational erosion"). Because we don't see anything out there right now, we know that these processes have been going on for a long time. Presumably, if we could see farther out, we would see the original uneroded disk material.

In this theory, Pluto is just one of many largish objects that used to exist in the inner belt. At some time during the early evolution of the solar system, the object that eventually became Pluto collided with one of its neighbors. This allowed Pluto to lose some energy and settle into its current stable orbit (the collision also gave the planet its small moon). Computer simulations indicate that only objects that had undergone such a collision could stay in the solar system as long as Pluto has. Thus Pluto seems different from the other planets because it *is* different. It's not really a planet but the last survivor of a large group of nonplanetary objects.

Over the next few years, astronomers using the Hubble Space Telescope will be making detailed observations of the regions of space outside the orbit of Pluto. If they find evidence that gravitational erosion is indeed the major factor determining the evolution of the outer solar system, then Pluto will no longer seem so strange. (I should note that as this book goes to press, there are already announcements of the discovery of a couple of hundred New Hampshire–sized objects out there, as predicted by this theory.)

In addition to casting light on the development of our own solar system, this new picture is going to be very important in the coming search for planets around other stars. Up until now, it has been assumed that other solar systems would more or less resemble our own. Frankly, after learning how sensitive the outer solar system seems to be to the influence of the large outer planets, I'm not so sure this assumption is valid. If the equivalent of Jupiter or Neptune in another system were a little bigger or smaller than ours are, or if their orbits were in slightly different places, it's quite possible that the outer part of another solar system could look very different from our own. It could, for example, have no Plutos or dozens of them. In either case, astronomers from planets around that sun might be astonished to see our system, with its single survivor of the original planetary disk.

4

Earth and Planetary Sciences

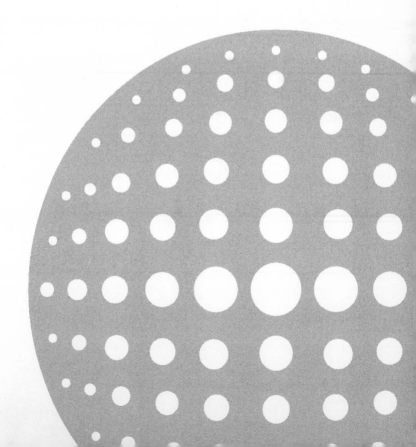

Was Malthus Right After All?

THE ENGLISH ECONOMIST Thomas Malthus (1766–1834) is recognized as the founder of the modern discipline of political economy. But for most of us, he is remembered as the gloomy predictor of the Malthusian dilemma. Here, from his *Essay on the Principle of Population as it affects the Future Improvement of Society, with Remarks on the Speculations of Mr. Godwin, M Condorcet, and Other Writers,* is the problem: "Population, when unchecked, increases in a geometrical ratio. Subsistence increases only in an arithmetical ratio. A slight acquaintance with numbers will show the immensity of the first power in comparison of the second."

In modern terms, the Malthusian dilemma can be stated simply: populations tend to grow exponentially, while the resources available in any ecosystem (up to and including the earth itself) have a fixed limit. Therefore populations will always exceed their resource base, resulting in widespread famine and death.

Since Malthus's time, there have been two classes of response to his arguments. Techno-optimists (of which I am one) argue that advances in technology will continually increase the resources available so that any ecosystem can support its human population even as that population grows. Malthusian doomsayers, on the other hand, continually predict imminent disaster. In the 1968 edition of *The Population Bomb,* ecologist Paul Ehrlich argued, "The battle to feed all of humanity is over. In the 1970's the world will undergo famines — hundreds of millions of people are going to starve to death in spite of any crash programs embarked upon now. At this late date, nothing can prevent a substantial increase in the world death rate."

Although there were some localized famines in the 1970s and 1980s, they tended to come about through the withholding of food as a political weapon (by Marxist dictators in Ethiopia, for example), rather than because populations reached a Malthusian limit. In fact, the 1970s will be remembered as the time of the "green revolution," a massive, technology-driven increase in world food production — the average daily calorie intake for people worldwide actually increased by 21 percent between 1965 and 1990, wheat yields per acre doubled, and rice yields went up 52 percent. Chalk one up for the techno-optimists!

But what about the next century? Can we go on feeding a growing human population until, as present models predict, it levels off at 8 billion in about 2050? That's where the debate centers now.

The Malthusians (who tend to be environmental scientists) have modernized their arguments somewhat and now argue that environmental damage to soils and croplands will prevent the increases in food production we've seen over the past decades. They argue that most good cropland is already under cultivation and that erosion is destroying much of what's left, especially in the third world. In their 1990 edition of *The Population Explosion*, Paul and Anne Ehrlich state: "Human numbers are on a collision course with massive famine. . . . If humanity fails to act, nature will end the population explosion for us — in very unpleasant ways."

Optimists (who tend to be economists and agricultural scientists) counter by saying that our agricultural system is nowhere near its limits. They refer to detailed studies indicating that four times as much land as is now farmed could be brought under cultivation (through the use of irrigation, for example), and that improved strains of plants will increase yields. World rice yields, for example, are only 20 percent of their theoretical limits. They also point to the usually ignored problem of wasted food (by some estimates, humans consume only about 60 percent of the food they grow) as an area in which significant progress could be made. They argue that the current depressed state of world cereal prices indicates that we already produce more food than we need.

Oddly enough, one of the looming environmental problems facing the world — the greenhouse effect — may actually play a positive role in food production. Climate models that predict global warming generally also predict an increase in world rainfall. More available fresh water, coupled with additional carbon dioxide in the air, will probably improve growing conditions considerably (the availability of carbon dioxide is what normally limits plant growth). In addition, if (as seems likely) the warming occurs as a slight increase in nighttime temperatures in the temperate zones, over the next century the growing season will be extended in northern latitudes, and large areas of Canada and Russia will be opened to agriculture.

This debate will go on, of course, but it seems to me that the techno-optimists are likely to win again. If they do, we can start to pay attention to the real question here — do we really want to live in a world with 8 billion people? There is, after all, more to life than food.

Is the Climate Getting Warmer?

 WE OFTEN SPEAK of "global warming" as if it were an established fact, yet there is a good deal of skepticism in the scientific community about whether the data we have reveal anything more than the normal variability of climate. I have to warn you that on this subject I am something of a skeptic, perhaps even a contrarian, so take what I say with a grain of salt. (And take what everyone else says the same way!)

The first thing to realize about the earth's climate and average temperature is that both are changing all the time. Over hundreds or thousands of years, changes of several degrees are common. (In this section I'll follow the general custom and report warming and cooling in degrees Celsius. For a rough conversion to the more familiar and accurate degrees Fahrenheit, multiply all my numbers by two.) Second, human beings have not been keeping temperature records for very long. The first thermometer was invented in 1602 (by Galileo, believe it or not), and the mercury thermometer didn't come into widespread use until about 1670. This leads us to one of the most difficult problems faced by scientists trying to discern trends in the earth's climate — because we don't have a very long record of temperatures, it is very hard to establish a base line to which any warming or cooling can be compared. In North America, for example, one of the best temperature records dates back to the early 1800s, when Thomas Jefferson started keeping records at his home, Monticello, in central Virginia. In Europe, records in several cities date back three hundred years.

To go beyond that, scientists have used all sorts of proxies for temperature. They have, for example, examined the dates of grape harvests in medieval France to estimate summer temperatures and rainfall, land deeds in Switzerland to track the advance and retreat of glaciers, and iceberg sightings in Iceland to estimate the climate in the Northern Hemisphere. The advent of the thermometer helped, of course, but raised its own problem — which temperature should we measure? Air temperatures (what we normally mean by "temperature") are measured at only a few places on earth (airports, for example). Water temperatures at the ocean surface are easy to measure by satellite. But both of these measurements present problems of interpretation. Airports, for example, are customarily built in rural areas that get covered with concrete. (Did you know that the designation for Chicago's O'Hare airport — ORD — refers to the apple orchards in which the airport was built?) Does a higher temperature twenty years later tell you about climate or about the effect of paving rural landscapes? Similarly, in the old days, sea-surface temperatures were sampled from buckets of water hauled up over the sides of boats. But after 1946, measurements were made of water coming into engine rooms. If the temperatures rose after 1946, do you blame the climate or the warmer environment of the engine rooms?

Over the past few years, this sort of debate has intensified as questions about global warming have moved from the scientific community into public discussions of the greenhouse effect. There are, in fact, two separate issues that often become confused in the public debate: (1) can we document a global warming, and (2) is that warming attributable to human actions?

The best long-range records, compiled primarily from sea-surface records over the past 140 years, indicate a warming of a little less than half a degree. Most of this warming occurred in two pulses — one in 1920, the other in 1977. In between these dates the sea-surface temperature was roughly constant, and, for ten years starting in the 1940s, actually dropped. (I remember this period well, because it provoked headlines about a "Coming Ice Age" in the Chicago newspapers.) The first of these pulses occurred well before we would expect any greenhouse warming based on increased carbon dioxide concentrations in the atmosphere. The second, unfortunately, occurred just before we began launching temperature-monitoring satellites.

Since 1979, when the satellite record began, there has been no discernible global warming, despite predictions from the computer models that the earth should be warming at a rate of about one-quarter degree per decade. In fact, the data actually show a slight cooling.

Faced with the problem of establishing the human influence on climate, though, the Intergovernmental Panel on Climate Change tried a new tack in 1996. Instead of looking at total warming, they looked at patterns of change, in particular the temperatures of the atmosphere at different altitudes and the pattern of warming and cooling in different parts of the world. They claim that they can just begin to see the imprint of human influence on climate. The big question, of course, is what will happen next. If, between now and the year 2000, global temperatures continue to fluctuate through normal ranges, we will have to question the validity of our computer models. If we see a rapid temperature increase, maybe even I will have to start taking global warming more seriously.

How Stable Is the Climate?

 WE ARE USED to having the weather change from day to day, but we expect the climate to be relatively stable. Minneapolis, we think, ought to have colder winters than Miami, no matter how much local weather patterns shift around. If the earth's climate changes, we expect it to happen over many thousands of years, so that living things can adapt. The conventional wisdom is that one of the great dangers of greenhouse warming, should it occur, is that average global temperatures would change rapidly and ecosystems would be unable to adjust before species are driven to extinction.

While the *amount* of expected greenhouse warming isn't unprecedented in the earth's history, the speed is. When the earth came out of the last Ice Ages, average global temperatures rose about 5 degrees Celsius. It has always been assumed, however, that this warming took several thousand years, allowing ample time for the earth's flora and fauna to adjust.

But these comfortable assumptions may not be justified. Most of the challenges to conventional wisdom on this subject are coming from studies of cores taken from the Greenland icecap. During all those years when glaciers were advancing and retreating, snow was falling in Greenland. Over time this snow was compacted, trapping whatever materials were in the air. Today scientists drill cores into the ice to retrieve samples formed up to 200,000 years ago (only the last 180,000 years or so have clear annual layers — think of them as analogous to tree rings). From these layers, information on temperatures (from measuring the abundance of certain chemical elements) and snowfall can be gleaned — in effect, they are repositories of past climates. If someone asks you, "Where are the snows of yesteryear?" you should answer, "In the Greenland ice cores."

The most striking feature of the data involves an event called the "Younger Dryas." After the ice from the last glaciers started melting, about 15,000 years ago, there was a general warming trend. Then, a little less than 12,000 years ago, the climate suddenly flipped back to cold. It stayed there for about 1,000 years, then suddenly warmed. And when I say suddenly, I mean just that — data on oxygen isotopes in ice and sea-floor cores indicate that the temperature in Greenland changed by some 7 degrees in as little as twenty years! This would be equivalent to changing the climate of Boston to that of Miami in two decades — a much more violent change than any contained in the greenhouse predictions. There are a lot of corroborating data for this finding, including studies of fossil plankton in deep ocean cores all around the Atlantic and cores taken from mountain glaciers in South America.

The main explanation for this phenomenon involves something known as the Great Conveyer Belt, which consists of ocean currents (like the Gulf Stream) that carry warm water north, where it cools and sinks. South-moving cold currents under the surface complete the circuit. The heat delivered to the Arctic by the Conveyer amounts to about 30 percent of the heat delivered by the sun. The theory is that the melting of the glaciers poured fresh water into the northern Atlantic, lowering the salinity of the water there. Fresh water doesn't sink, so its presence shut down the Conveyer. The result: an abrupt change of temperature at the poles. As the fresh water spread out, the salinity increased and the Conveyer started up again.

In one of those examples of serendipity that occur now and then in the history of science, information on these sudden climate changes has caused theorists to link the new data with some old (and rather puzzling) data obtained from cores drilled into the north Atlantic sea floor. These data seemed to indicate that every 7,000 to 12,000 years, a layer of continental rocks was being dumped in the middle of the Atlantic Ocean. The only apparent explanation for this anomaly is that these rocks were locked into the bottom of icebergs and rafted out into the ocean, to be dumped wherever the iceberg melted. But why should these raftings occur so regularly?

One possible answer to this question rests on the nature of Canadian geology. Most of the rocks in northeastern Canada are hard and crystalline. Around Hudson Bay, however, the rocks are softer. The theory: as the ice built up to more than 10,000 feet on Hudson Bay, the rock underneath fractured and, mixing with melted water from the ice sheet, turned into a layer the consistency of toothpaste. When this happened, the ice covering Hudson Bay slid out over its "toothpaste" into the North Atlantic, creating an iceberg armada that dumped rocks on the ocean floor and supplied enough fresh water to shut down the Great Conveyor Belt.

Further studies of the Greenland ice cores and the fossil record in the Atlantic will show whether this particular theory, or some variation on it, explains the observed rapid climate changes in the earth's past. The bottom line, though, is that our planet has a much more violently variable past than we have believed, and the fact that the climate has been more or less stable for the last 8,000 years is no guarantee that it will remain so in the future.

The Case of the Missing Carbon

 WHENEVER YOU BREATHE or drive a car or cook on a gas stove, you are producing carbon dioxide, a molecule made from one carbon atom linked to two oxygens. Carbon dioxide isn't a pollutant — it and water are the natural outcome of combustion, whether in a man-made machine or a living cell. Plants need atmospheric carbon dioxide to grow, and the earth's atmosphere and oceans have always contained a certain amount of it. Our attention is focused on it these days for two reasons: (1) human activities, such as burning fossil fuels and cutting down the rain forests, are adding carbon dioxide to the atmosphere in measurable amounts, and (2) carbon dioxide reflects outgoing radiation back toward the earth, and is therefore one of the key contributors to the greenhouse effect and possible global warming.

We know that some of the carbon dioxide you created when you drove to work this morning will eventually wind up in the ocean, some will be taken up by plants, and some will remain in the atmosphere to contribute to the greenhouse effect. But one of the persistent scientific embarrassments of the past several decades is that when we try to balance the books — add up all the carbon entering the atmosphere and subtract everything that goes into a known reservoir or stays in the air — we find that hundreds of millions of tons of carbon are not accounted for. They seem to have vanished completely somewhere between the smokestack and the ecosystem.

Here's how a typical accounting might go: deforestation and the burning of fossil fuels add about 6.5 to 7 billion tons of carbon to the atmosphere each year. At the end of the year, however, only about 3 billion tons remain in the air. If you ask oceanographers how much carbon goes into the ocean, they will tell you about 2 billion tons, which means that the remainder is being soaked up by land plants. The best surveys of these plants, though, have always indicated that they were at least a half billion tons a year short of absorbing the balance of the carbon. The stuff that's unaccounted for has become known in scientific circles as the "missing carbon."

Over the years people have steadily chipped away at this problem, looking for places where previously unnoticed carbon might be stored in large amounts, hoping that each new discovery would account for it all. My favorite explanation was hatched during the early 1980s, when what my preadolescent son called the "fish poop" theory was riding high. In this theory, the missing carbon was supposed to reside in what was delicately called "organic detritus" on the ocean floors.

Since then, however, our understanding of the global carbon balance has become much more detailed and precise. We know that most of the carbon in the biosphere is in the ocean (39,000 billion tons vs. only 600 billion tons in the atmosphere); most of it is in the form of dissolved carbon dioxide, as if the seas were one vast (and slightly flat) carbonated drink. Carbon from the atmosphere moves into the ocean, and carbon from the ocean routinely moves into the atmosphere. In the earth's temperate zones there is a net flow of carbon into the sea; in the tropics there is a net flow into the atmosphere. Worldwide the ocean absorbs 2 billion tons more carbon per year. It seems unlikely to me that the oceans will be found to be hiding the missing carbon.

Over the past few years, a number of surveys of terrestrial plants have identified at least some possible places where the missing carbon might be hiding. For one thing, although it is true that we are reducing the amount of rain forest in the tropics, forests in the temperate zones are flourishing. If you hike through the oaks and maples from New England to the Carolinas, you constantly stumble on old fences and fireplaces — reminders that a century ago this land was open fields. Some estimates indicate that the amount of carbon taken up by new temperate forests may well exceed the amount returned to the atmosphere by cutting down forests in the tropics. Furthermore, studies of the root systems of grasses in the South American savannas indicate that they could remove up to half a billion tons of carbon from the air annually. Thus, the pastures that replace the rain forest may turn out to be significant absorbers of carbon.

Today scientists seeking the missing carbon are focusing on terrestrial plants. My guess is that people will start to do more detailed studies of temperate forests, Arctic tundras, and grasslands to see where the carbon has gone, and that as time goes by the books will come more and more into balance. This is very important from a policy viewpoint, because unless you know where the carbon comes from and where it goes, it's hard to know whether or not we can lessen the greenhouse effect by reducing the amount of carbon we put into the air by burning fossil fuels.

Are We Going to Lose the Ozone Layer?

 PROBABLY NOT, because the response to the threat to the ozone layer is the best example I know of science and public policy working together.

Ozone is a molecule made from three atoms of oxygen. Most of it resides in a region of the atmosphere fifteen or twenty miles above the earth's surface, called the ozone layer. This is where most of the ultraviolet radiation from the sun is absorbed, but it is also the region that was threatened by the use of chlorofluorocarbons (CFCs). (I should point out that even in the ozone layer, ozone is a tiny fraction of the gases in the atmosphere.)

I think of the period before 1985 as the "spray-can era" as far as the ozone layer is concerned. In that period we knew that CFCs entering the atmosphere could, because of their extreme stability, make their way up into the ozone layer, where the action of sunlight on them releases chlorine. A small fraction of the chlorine acts as a catalyst to break up ozone molecules. It was calculated that the spray-can effect would deplete the ozone layer at the rate of about 5 percent per century. Because of this finding, Canada, Sweden, and the United States banned the use of CFCs in spray cans in 1985.

The mid-1980s also marked the discovery of the Antarctic ozone hole, the severe annual depletion of ozone above that continent during the Antarctic spring. At first, there was a great deal of debate about the ozone hole, particularly about what caused it. It turns out that the ozone hole appears over Antarctica because it is the coldest place on the planet. During the long Antarctic night, polar stratospheric clouds (PSC), made of tiny ice crystals, form high up in the ozone layer. Most chlorine in the atmosphere is taken up into molecules of hydrochloric acid which, although it may contribute to acid rain, has no effect on ozone. In the PSCs, however, hydrochloric acid deposits on the surface of the ice crystals, and chemical reactions take place that form chlorine oxide (ClO). Chlorine oxide does not break up ozone molecules by itself, but in the presence of sunlight it dissociates to form ordinary chlorine, which can disrupt the ozone.

During the Antarctic winter, then, the supply of chlorine oxide builds up and there is no sunlight to get rid of it. When the sun returns in the spring, two things happen: first, the PSCs disappear, and second, the chlorine oxide breaks up, freeing the chlorine. The result is a burst of ozone destruction that goes on until the chlorine chemistry stabilizes. This is what causes the ozone hole.

Recent studies have shown a similar, but much milder, effect in the Arctic. The Arctic is not cold enough to have PSCs, so the destruction of the ozone is much less. While the Antarctic ozone hole might show depletions of 30 and 40 percent, the Arctic ozone hole very rarely goes above 10 percent.

Recent studies have also shown a somewhat smaller degradation of the ozone layer over the mid-latitudes, where the depletions are seen at various times. There are arguments about what could cause such an effect. One is that large volcanic eruptions, which occur every decade or so, put particles into the air that play the same role in the mid-latitudes that PSCs play in Antarctica. Another possibility is that air from the Arctic and Antarctic moves into the mid-latitudes, carrying chlorine with it and thus reducing the ozone in those regions.

In 1987 the industrialized nations of the world met in Montreal and agreed to phase out production of CFCs. This has been an extremely successful program — so much so that the nations have met periodically since then, most recently in Copenhagen in 1992, to tighten the standards. The Copenhagen Accord calls for complete elimination of the manufacture of CFCs by 1996.

So what will happen to the ozone layer, assuming that the Copenhagen Accord goes forward? Calculations indicate that chlorine levels in the atmosphere will return to normal in about the year 2030.

This is an amazing success story. In the first place, scientists identified and diagnosed the problem in a timely fashion. Then governments listened to the scientists and acted to eliminate that problem. Finally, it appears that because of this combined action the problem is going to go away. How could the system work any better than that?

Could We — Should We — Modify the Climate?

 AT THIS POINT in history it is clear that, through smokestack emissions and other such processes, we can inadvertently alter the earth's climate. Can we, then, alter it *advertently?* Is there a way to counter the bad effects of human activities on the climate or even to change the climate to be more to our liking?

The answer to this question depends on two factors: (1) does our present technology allow us to make large-scale changes in the climate, and (2) if it does, what political, legal, and ethical considerations are involved?

Most of the thinking that's gone into this question has had to do with the greenhouse effect. Here are some numbers to set the stage: above the atmosphere, the sun pours down energy at the rate of about 1,500 watts per square meter — enough to run a toaster and four light bulbs continuously. Doubling the carbon dioxide in the atmosphere is equivalent to increasing this figure by about 4 watts per square meter. This is the approximate level of greenhouse warming people expect by the end of the next century. However, a large volcanic eruption, like that of Mount Pinatubo in 1993, puts sulfur compounds into the stratosphere and increases the amount of sunlight reflected away from the earth and back into space. The effect of the Pinatubo eruption was equivalent to lowering the incoming energy by about that same 4 watts per square meter; it caused a three-year global cooling.

Because the climate is so sensitive to changes in energy balance, one class of climate-modification suggestions would operate to reflect more sunlight into space. Space mirrors, space dust, fleets of balloons, stratospheric dust spread by 747's, and increased cloud seeding have all been suggested as ways to counteract the greenhouse effect. No large-scale experiments have been done or are planned in this area, but large volcanic eruptions have supplied us with natural experiments. Pinatubo, for example, put about a million tons of sulfur compounds into the atmosphere. Mixing some of these compounds into ordinary jet fuels could keep enough sulfur up there to cancel the greenhouse effect. In other words, it is clearly within our capability to modify the climate in this way.

The question, of course, is what else the sulfur compounds would do. There's the rub. We don't really know enough yet to predict the climatic effects of greenhouse gases with any degree of confidence. By the same token, we don't know enough to predict the side effects of this sort of climate modification scheme. It will be a good ten years before our theories are good enough to make such predictions with any confidence, and that seems to be the main technical problem with all ideas based on reflecting sunlight away from the earth.

Another set of schemes has to do with removing carbon dioxide from the atmosphere. Actual testing is being done on one, which works like this: in some regions of the oceans the growth of photosynthetic plankton is limited by the scarcity of iron. What if we just dumped iron into the sea? The plankton would grow, pull carbon dioxide from the air, and (hopefully) take it down into the deep ocean when they die. Again we have a climate alteration scheme that, on the surface, seems plausible.

For this case, we also have a recent experiment. A group of researchers from the Moss Landing Marine Research Lab dumped iron filings into the ocean near the Galápagos Islands to see how the scheme worked in the open ocean. They found that the iron did, indeed, promote the growth of phytoplankton. In fact, the reasearchers found that the number of plankton in their patch increased tenfold. The ocean, quite literally, turned from clear blue to soupy green for miles around the ship. "It was like sailing into a duck pond" was how one participant put it. All told, the seven-day trial removed about 100 tons of carbon from the atmosphere.

But as with the atmospheric schemes, we don't yet know enough about ocean fertilization to say with confidence what the side effects might be. One suggestion: when the fertilization stops and the dead plankton sink to lower depths, their decomposition might break up the nitrogen compounds whose upwelling supports many of the earth's fisheries. Only more research will answer questions like this.

The political problems associated with climate modification are actually more difficult than the technical ones. One clear difficulty is that at present there is no international body that can authorize large-scale climate modifications. Some deliberative bodies have been set up by treaties — the Rio Treaty, for example, or the Law of the Sea — but they can't make decisions. Furthermore, to a litigious nation like ours the questions of liability are daunting. A lawyer I know summed up the situation best. "Will we get sued if we try this stuff?" I asked.

"Sure you will." he said. "But then, you'll get sued if you don't try it, too."

How Clean Is Clean,
and How Much Is It Worth?

OUR ABILITY to detect small amounts of chemicals has increased rapidly over the past two decades, a fact that has enormous consequences for the environmental debate. An industrial site that would have been considered "clean" ten years ago might not be considered "clean" now, even if it has not changed. Consequently, our perception of pollution is changing.

Some background: if you dissolve a cube of bouillon in a quart of water, the concentration is about one part per thousand, and the human eye can detect the "pollution." If you dissolve a piece of bouillon the size of a grain of sand in the quart of water, the concentration is one part per million (ppm) and you need instruments to see it. If you dissolve a barely visible speck of dust in the quart, the concentration is one part per billion (ppb). Today we can routinely detect pollution at this level and beyond.

In a purely scientific discussion, this fact raises no particular difficulty. The problem is that our legal and regulatory systems, developed when we did not have such detection capability, do not seem able to cope with the new technology. If a law requires that no measurable amount of any substance known to cause cancer in laboratory animals can be present in food (or, by extension, in the environment), those laws hold even in cases where the amount is big enough to measure but too small to constitute a health hazard.

The outcomes of some situations are truly bizarre. For example, standards at the Nevada nuclear test sites require that radiation levels inside the test area actually be less than the natural background levels on the other side of the fence. In other words, because of our ability to detect pollutants, we are sometimes required to make a site "cleaner" than nature itself!

I expect that several aspects of the debate over "how clean is clean" will continue for many years. One conflict concerns our ability to detect small amounts of a material and our sense of the amount that poses a health or safety risk. This debate often goes under the name of "risk assessment." The idea is that the level of a particular substance that we allow in the environment ought to be related to the risk posed by the substance. In some cases, this might mean that we would tolerate levels considerably higher than the minimum that we can detect instead of, as at present, demanding removal. There are obvious political problems with changing laws in this way, but it would have the enormous advantage of freeing our environmental standards from being held hostage by our technical vituosity.

A second strand of the "how clean is clean" debate will be over whether environmental standards should be set according to anticipated use. In cleaning up an old factory site, for example, should we have less stringent standards if the site is to become an oil refinery than if it is to be a day care center? Asking the question in this way makes the answer obvious, but the law today does not allow this sort of distinction.

This is an important topic for three reasons: first, removing (say) 95 percent of a contaminant is often easy and relatively cheap, while getting rid of the remaining 5 percent can be enormously expensive. Thus, a system that set one level of cleanup to clear a site for use as a factory or warehouse but required much more extensive work for residential use would recognize economic realities. It would also avoid the altogether too common situations in which people have gone into mildly contaminated sites and bulldozed acres and acres of perfectly fine natural ecosystems and dumped all the slightly contaminated soil into a landfill, simply because that was the easiest way to meet environmental standards.

Second, in many cases there is simply no way to remove all the pollution from a site. In contaminated soil, for example, some of the pollutants are in the spaces between the soil particles, where they can be washed out easily, but some may be inside the soil grains themselves. This means that you can wash the soil, then come back a year later to find that the soil particles, like a squeezed-out sponge, have diffused more of the pollutant into the environment. We don't yet have the technology to deal with this problem.

The final reason I find flexible cleanup standards so attractive is that the current laws have done enormous damage to an environment I care about deeply — America's central cities. Under present law, the minute you sign your name to a deed, you are liable for all expenses incurred in removing pollution from that site, even that required by the future detection of now undetectable levels of any substance. Consequently, many so-called brown-field sites in our cities lie vacant and unused because no lender or developer is willing to take on the risk associated with the environmental cleanup. If there was ever a case of the best being the enemy of the good, this is it!

Will the Antarctic Ice Sheet Melt?

 YOU MAY HAVE seen the poster during the last big scare about global warming — the one with the Statue of Liberty up to her neck in water. The implicit message of the poster was that global warming would melt all the world's glaciers, leaving the statue a victim of rising sea levels. Although no responsible scientist today would consider such a turn of events likely, there is a grain of truth in the poster image. A small percentage of the world's water is locked up in ice and glaciers — mainly in Antarctica and Greenland, but scattered around the world's mountains as well. If all of that ice melted, sea levels would probably rise considerably, perhaps even the couple of hundred feet it would take to put the Statue of Liberty in peril.

Since most of the world's ice is located in Antarctica, most of the concern about rising sea levels caused by melting ice has been focused on that continent. And although research has given us some rather surprising insights into the nature of large bodies of ice, it has also touched off an ongoing scientific debate on the question of how much, if any, of the Antarctic ice sheet is likely to melt anytime soon. Oddly enough, very little of that debate is concerned with the greenhouse effect — the attention focuses on the dynamics of glaciers instead.

Massive glaciers are not just big blocks of ice but dynamic systems. Snow falling at higher altitudes packs down, accumulates over a period of years, and flows downhill like a slowly moving river. At lower altitudes, the front face of the glacier melts.

Antarctica has two different ice sheets with very different

properties. On the eastern part of the continent, the ice is underlain by solid ground — mountain ranges and plains. Here the ice piles up several miles high and moves very little. Most scientists regard the East Antarctic ice sheet as pretty stable, and it's unlikely to change much on time scales relevant to human beings.

The West Antarctic ice sheet, though, is a different story. For one thing, a good part of it isn't on land but floats out over the ocean. For another, its motion is quite complicated. Over land, the temperature and pressure under the sheet are high enough to melt some of the ice and turn the underlying rock and soil into a material about the consistency of toothpaste. On this lubricated surface, the ice "floats" toward the sea. In addition, if you fly over the sheet, you can see that the broad expanses of smooth ice are traversed by regions of broken ice and crevasses. These are ice streams that flow through the slower-moving sheet and account for a good deal of the ice the sheet dumps into the ocean. The ice streams can move at surprising speeds — several miles per year are not uncommon. Their motion is also erratic. They speed up, slow down, and sometimes just stop flowing altogether.

As you might guess, the behavior of the West Antarctic ice sheet is extremely difficult to predict. In general, it flows into the ocean, where it produces icebergs that float out and melt. The details work like this: ice flows faster when it floats on the ocean than when it drags over land, so the part of the sheet that is over water pulls on the rest. As a result, the point at which the ice begins to float on the underlying water — the so-called grounding line — moves farther inland. If no other effect were operating, the entire West Antarctic ice sheet would be pulled into the sea and melt, raising world sea levels by about eighteen feet. That isn't enough to inundate the Statue of Liberty, but it would certainly cause big problems for coastal cities.

Fortunately, there is another process that affects the location of the grounding line. As the ice sheet moves toward the sea, the lubricant at the base gets dumped into the ocean as well, building up a deltalike deposit of "toothpaste." This tends to move the grounding line seaward. At the moment, the state of the West Antarctic ice sheet is determined by the delicate interplay between these two effects — one tending to pull the sheet into the sea, the other tending to stabilize it.

The question is whether and how long this equilibrium can last. And this is where the controversy starts, because our understanding of this process is still quite primitive. The conventional wisdom is that the ice sheets formed, more or less as they are now, when the Antarctic continent moved to its present position over the South Pole about 14 million years ago. However, some evidence (mainly of fossils found in drilling cores) suggests that in recent geological times — perhaps no more than 400,000 years ago — the West Antarctic sheet disappeared completely. This suggests that the equilibrium of the ice sheet may be as delicate as it appears to be, and that a small change in conditions might be enough to start it melting. This point of view got unexpected support in 1993, when scientists found a Mount Fuji–sized volcano completely buried under the West Antarctic sheet, along with evidence that the volcano is still active. The Doomsday scenario: an eruption under the ice triggers a seaward surge that dumps the whole sheet into the ocean.

How Does the Earth's Mantle Move?

 THE MODERN PICTURE of the earth holds that the motion of the continents is driven by heat generated by radioactivity deep inside the planet. This heat rises to the surface through convection (the same process you see in a pot of boiling water, in which hot water rises from the bottom, then cools off and sinks to repeat the cycle). Over hundreds of millions of years, rocks deep underground go through the same cycle, with the plates and continents riding along the top of the moving rock like bubbles on boiling water.

But although there is superficial agreement among scientists to this point, the arguments start when we get into details. In particular, there is continuing debate about how deep the convection process extends into the earth's interior.

Some background: after the earth was first formed, it went through a period of melting all the way through. Heavy materials sank to the center, and the rest of the planet acquired a layered structure. Outside of the nickel-iron core and extending to within a few miles of the surface is a 2,000-mile thick layer of heavy rock called the mantle. It is the convection of this rock that moves the continents.

Scientists cannot study the interior of the earth directly, but they can observe the arrival of seismic waves at observatories on the surface. Every time a major earthquake occurs, waves travel through the earth, and by measuring how long it takes them to get to different points on the surface, scientists can build up an amazingly detailed picture of the earth's interior. This technique can be compared to the medical procedure of a

CAT scan, in which pictures of the body are built up by measuring the absorption of X-rays.

Seismic studies reveal one outstanding fact about the mantle: a layer about 400 miles down seems to mark a transition between the upper and lower mantle. Seismic waves apparently speed up when they pass down through this transitional layer, which can be explained only by assuming that the atoms in the lower mantle are arranged differently from those in the upper mantle.

For the last thirty years a controversy has raged among geologists over the question of whether the roiling convection that drives the continents extends down only to this layer, or whether it goes all the way to the bottom of the mantle. As often happens with long-standing controversies in the sciences, the basic problem is that there are two sets of data, from two different disciplines, that cannot be reconciled.

On one side, geochemists point to the presence on the earth's surface of a number of chemical elements whose abundance on the surface seems out of proportion to their abundance in the upper mantle. The geochemists argue that this kind of imbalance can occur only if material from deeper in the mantle is brought up occasionally, which would happen if convection mixed the two mantle layers. On the other side, seismologists argue that they see no evidence for such mixing in their "CAT scans" of the earth.

This debate is still going on at full force, but it may be resolved within the next few years. New information, both in improved data analysis and in more precise computer modeling of the earth, is pointing the way to compromise solutions.

It turns out that seismic waves travel faster in cold rock than in warm, which gives seismologists a way of taking the temperature of rocks deep within the earth. The problem of doing this at great depth has always been that as waves move toward the surface, they are distorted by inhomogeneities on the rock, much as light waves shimmer as they pass through the heated air above a parking lot on a hot day. One recent analysis, which accumulated enough data to address this problem, showed a plume of cold rock sinking all the way through the mantle off the West Coast of the Americas. This was the first direct evidence of a breach of the mid-mantle barrier.

On the theoretical front, a number of ideas have been proposed that might accommodate both points of view. In one scheme, cold rock sinks to the boundary, where it starts to pool. When enough has accumulated, it breaks through into the lower mantle, and lower mantle material squirts up through the breach. In another model, periods like the present, when convection is mainly confined to the upper mantle, are interspersed with periods when plumes of heated material rise from the bottom of the mantle, like bubbles in a pot of heated water before boiling starts. In both of these schemes, there is plenty of opportunity for mixing the two layers of the mantle, but that mixing doesn't go on all the time. My guess is that when we get more data, some model like this will turn out to be correct.

Why Does the Earth's Magnetic Field Reverse Itself?

 IF YOU'VE EVER used a compass, you are probably familiar with the fact that the earth behaves like a giant magnet, with its north pole in Canada and its south pole in Antarctica. A compass needle, which is a tiny magnet, aligns itself in a north-south direction because of its interaction with this terrestrial magnet. What you may not know is that compass needles haven't always pointed north.

Actually, it wasn't until well into the twentieth century that scientists figured out why the earth had a magnetic field in the first place. Physicists had known for a long time that running an electric current through a wire would produce a magnetic field — this is the basic operating principle of both the electric generator and the electric motor. The first thought was that perhaps the rotation of the earth produces an electric current in the interior and that this electric current in turn produces the magnetic field. But the earth as a whole has no net electrical charge, so its rotation wouldn't be an electric current.

The currently accepted theory for the origin of the earth's magnetic field involves the behavior of the planet's liquid core. About 2,000 miles beneath your feet, the temperature approaches 6000 degrees C (approximately the temperature of the surface of the sun) and the pressure is more than a million times that of the atmosphere. Under these conditions, the iron and nickel in the core are liquid. Farther in, under even greater pressure, the atoms are squeezed together into a solid, but a stretch of the earth's interior some 1,500 miles deep is liquid. Think of this as a roiling, steaming, turbulent ocean bounded by solids above and below and you have a pretty good picture of it.

If a liquid metal like this rotates in a small preexisting magnetic field, then loose electrical charges (both positive and negative) will start to flow, creating an electric current. This current will build up the magnetic field, which will then produce a stronger current, which will make a stronger field, and so on. In this way even a microscopic magnetic field will become stronger, and most scientists now believe that the magnetic fields of Earth and the sun were built this way. Bodies without a liquid core (like the moon and Mars) shouldn't (and don't) have such a field.

This would be fine except for the rather startling fact that the earth's magnetic field doesn't always point in the same direction. Every once in a while it seems to flip. There have been periods when the north pole was in Antarctica while the south pole was in Greenland or thereabouts. The most recent such period ended about 750,000 years ago.

You might wonder how we could possibly know what direction a compass needle pointed in so long ago. The answer has to do with the fact that outflows of molten rock, as in a volcanic eruption, have been fairly common in the earth's history. Molten rock often contains tiny grains of iron ore, each of which can be thought of as a tiny compass needle. In a liquid, the grains are free to rotate and align themselves so that they point toward wherever the north pole happens to be at the time of the outflow. Once on the earth's surface, the rock solidifies and the orientation of the grains of iron is locked into the crystalline structure. Those grains will continue to point in the same direction, regardless of where the north pole is later. The rocks, in other words, "remember" where the north pole was when they were formed. The study of those memories is a large part of the field of geology known as paleomagnetism.

Scientists in the 1960s and 1970s discovered systematic patterns in the alignment of "compasses" in old rocks. Sometimes those grains pointed north, sometimes south. This was what led to the notion that the earth's magnetic field reverses itself. Today, we can document more than three hundred such reversals of the field.

It's really hard to see how a simple model invoking the rotation of the liquid core could explain this behavior, especially since the timing of the reversals seems to be erratic. There are long periods with no reversals at all, followed by periods of rapid reversals.

Most of the theories that attempt to provide the beginnings of an explanation cite some kind of turbulent motion in the liquid cores — storms on the earth's interior ocean. The idea is that at any one time some features of the flow would put the north pole where it is now and other features would put it in Antarctica. The actual magnetic field results from the delicate interplay between these features, and very small changes can tip the balance and lead to a reversal. Such theories seem very promising, but to my mind, the question of why the earth's magnetic field reverses itself remains the greatest unanswered question in the geological sciences.

How Well Can We Predict Volcanic Eruptions?

 ACCURATE PREDICTION of natural disasters has always been a goal of science. Today we can predict the behavior of hurricanes very well. Typically we have several days of warning, and even a major storm like Hurricane Andrew causes relatively little loss of life because people can be evacuated from its path. We do less well at predicting earthquakes, for present technologies allow us to forecast only general probabilities over long periods of time. Prediction of volcanic eruptions falls between these two extremes. Over the past decade we have found out how to predict eruptions in time to save lives — and this ability will get much better in the near future.

A few numbers tell the story. In 1985, Mount Nevado del Ruiz in Colombia erupted without warning, killing more than 20,000 people. In 1991, only six years later, the time of the eruption of Mount Pinatubo in the Philippines was predicted accurately enough to evacuate most of the people living in the area. The result: only a few hundred deaths, rather than thousands. In a similar event in 1994, 20,000 people were moved without loss of life before an eruption in Papua New Guinea. For reference, the destruction of Pompeii by an eruption of Mount Vesuvius in A.D. 79 is estimated to have killed 20,000 people; the eruption of Mount Tambora in Indonesia in 1815 (the largest eruption in modern times), almost 100,000.

A volcano is a place where hot magma from the earth's interior comes to the surface. The process starts when molten, relatively light magma deep in the earth starts to rise toward the surface, much as a block of wood released underwater will pop to the surface. As the magma rises, the pressure drops and dissolved gases start to form bubbles. Contrary to general belief, not all eruptions are explosive. If the magma isn't very viscous and contains few bubbles, lava will just flow out at the surface. But if the magma is viscous and contains a lot of dissolved gases, the result is the classic explosive eruption.

Typically, a volcanic eruption is preceded by all sorts of telltale signs — increased tremors in the ground, small eruptions, and the release of ash and gas (particularly gases containing sulfur compounds). Traditionally, volcanologists made their predictions from such indicators and a generous amount of "seat of the pants" calculation. Today two pieces of modern technology — the computer and the satellite — have been instrumental in enhancing our ability to (1) predict when eruptions will occur, and (2) predict what will happen when they do.

Scientists in the field can now perform instant analyses of new data on their personal computers — analyses that used to be done in distant laboratories on huge mainframe computers. The time lost in getting data to a mainframe, of course, can be crucial in making day-to-day predictions as a volcano nears eruption. In addition, it is now possible to place instruments on a volcano, then monitor them from a satellite. This means that as eruption approaches crucial measurements can be made without taking the risk of sending teams up on the volcano. It was a combination of satellite monitoring and real-time analysis with personal computers that allowed volcanologists to make accurate determinations of the eruption of Pinatubo.

In the future, volcanologists will have another tool to improve their monitoring: the Global Positioning System, or GPS (see page 67). As an eruption approaches, strains and pressures within the earth cause small bulges to appear at places on the flanks of the mountain. If sensors are put on the mountain before the bulging starts, the GPS can determine their positions and trace the gradual rising or sinking of the ground.

Computers can also help governments plan what to do when the volcano finally erupts. The primary cause of death in the Nevado del Ruiz eruption, for example, was a series of mudslides down the mountain. Had people known where those slides would occur, they could have moved to high ridges and avoided them. Today scientists are plotting the probable paths of mudslides, lava flows, and ash clouds (which can shut down the engines of aircraft that have the misfortune to fly through them) in detailed simulations of eruptions of specific volcanoes. The first such simulations were done in 1994 in anticipation of the eruption of Mount Popocatépetl in Mexico. The computer simulations worked out the most likely paths for mudslides, giving officials some much-needed guidance in their evacuation planning.

Today programs involving international agencies and organizations like the U.S. Geological Survey rush teams of scientists to a volcano that seems to be waking up after a period of dormancy. These teams monitor the volcano and start making predictions about eruption dates and consequences. My sense is that it won't be long before volcanic eruptions join hurricanes as one of those scourges that, if they can't be prevented, can at least be managed without extensive loss of human life.

Will We Be Able to Predict Earthquakes?

HAVING LIVED for some years in California, I can testify from firsthand experience that nothing is quite so frightening as an earthquake. Even though the ones I experienced were minor tremors, hardly worthy of notice in the national media, they were enough to give me a terrifying sense of what "The Big One" will be like when it comes.

Although they can (and do occasionally) occur elsewhere on the planet, most earthquakes take place in regions like the legendary San Andreas Fault, where the tectonic plates that make up the earth's surface move with respect to each other. The Pacific plate, which includes some of the coastal regions of southern California, is moving northward with respect to the North American plate at the rate of a couple of inches per year. The effect is similar to what happens when you bend a pencil. For a while the pencil holds, perhaps deforming a bit. Then all at once, it snaps. And while it's easy to say that the pencil will eventually snap and that the plates will eventually slip past each other and produce an earthquake, saying exactly *when* either event will happen is very difficult. This is the problem of earthquake prediction. We simply don't understand the process well enough to predict when the plates will let go.

Before getting into prediction, let me deal with one side issue. We are used to talking about the magnitude of earthquakes in terms of the Richter scale, devised in 1935 by the Caltech geologist Charles Richter, which measures the ground motion an earthquake creates. In fact, geologists no longer use the Richter scale; they use scales that allow them to estimate the total energy released by rock movement. They usually glide over this point by talking about earthquake "magnitude" without mentioning Richter. Most of the quakes you read about measure between 6 and 8 on these magnitude scales.

There are a lot of folktales connected to earthquake prediction. Perhaps the most widespread is that animals can sense that one is coming. The best-documented case of animal prediction occurred in Taijin, China, on July 18, 1969. Zookeepers observed all sorts of weird behavior — swans avoiding the water, pandas screaming, snakes refusing to go into their holes — just before a major earthquake. Scientists speculate that the straining rocks produce changes in the electric field near the earth, and that these changes somehow upset the nervous systems of some animals. Unfortunately, the animal behavior theory hasn't been borne out well since; there are many examples of quakes with no abnormal animal behavior beforehand.

By far the greatest scientific effort in prediction has gone into measuring changes at the earth's surface in earthquake zones. The idea is that as a fault gets ready to slip, there ought to be some sort of detectable change, such as surface bulging or rock strain. If we could specify some precursor of earthquakes similar to the pre-eruption rumblings of a volcano, we would have a prediction scheme. One step on this path was completed in 1995, when scientists succeeded in identifying some points where strain is building up along a fault in Bear Valley, California.

The most ambitious attempt to measure earthquake indicators has been going on near the town of Parkdale, California. The town is on a part of the fault that produced earthquakes in 1881, 1901, 1922, 1934, and 1966. Based on past performance, an earthquake could be expected again between 1988 and 1992, and in anticipation of this event instruments to measure ground tilt, water level, strain in rocks, and small precursor earthquakes were installed and monitored continuously. As of this writing, however, Murphy's Law has been shown to apply to earthquakes as well as to everything else — the expected quake hasn't shown up, although a lot of data have been collected.

One area that is just beginning to be explored is the use of the Global Positioning System to gather data for earthquake prediction. With the proper instruments, this system can detect shifts of ground position of less than an inch, so it should be possible to monitor many sites much more cheaply and easily than has been done in the past. The hope is that the science of earthquake prediction will someday come up to the level of volcanology, which seems to be getting pretty good at predicting eruptions.

But one aspect of earthquake prediction hasn't gotten much attention. Suppose that after making measurements of strain you predict that there is a 75 percent probability of a magnitude-8 earthquake occurring in the Los Angeles basin sometime during the next three months. What should you do with this information? If you make a public announcement, people might leave the area, which could easily cost billions of dollars in lost business opportunities alone. And what if, after all that, the quake doesn't come? Can you imagine the lawsuits?

But suppose you keep the information to yourself and the quake does happen. Can you imagine the reaction when people find out that the information has been available in advance? My own sense is that earthquake prediction may well be an area in which we don't want to make progress.

Why Are There Earthquakes in Missouri?

EVERYONE KNOWS that there are earthquakes in San Francisco because the city sits on top of the San Andreas Fault. The earthquakes result from the grinding, slipping, jerking motion as the Pacific plate churns slowly northward with respect to the North American plate. But did you know that the biggest earthquakes that ever hit the continental United States took place not in California but in New Madrid, Missouri, in 1811 and 1812? And that the third largest earthquake (after the 1906 quake that leveled San Francisco) took place in Charleston, South Carolina, in 1886? An enduring question that scientists are just starting to address is how earthquakes can occur far from the boundaries of tectonic plates.

A word of clarification: the term "tectonic plate" is not the same as "continent." Tectonic plates are the large blocks of material that make up the surface of the earth. The movements of the plates cause the drift of the continents, which ride along as passengers on the plates. Plates may or may not have continents on them, and the boundaries of the plates may or may not coincide with the continental boundaries. The North American plate, for example, extends from the West Coast of the United States (where part of the boundary is the San Andreas Fault) to the great mountain chain that constitutes the sea floor in the middle of the Atlantic Ocean. Both New Madrid and Charleston are far from the boundaries of the plate, even though one is on the coast. (Incidentally, as a Midwesterner, I have to point out that natives pronounce "Madrid" with the accent on the first syllable and the "a" as in "cat".)

As rudimentary as our predictive ability is for earthquakes along tectonic plate boundaries, it far exceeds our ability to predict (or even to understand) mid-plate events. This has important consequences, because regions where earthquakes are not expected have not adopted building codes that will protect life and property should one occur. In California, new construction has had to meet increasingly rigorous standards over the past half century, as engineers and scientists have accumulated knowledge of how structures react to the movements of the earth. Such standards have not been required in the Midwest or on the East Coast, so everything built in these areas — skyscrapers, highway bridges, tunnels — is at risk.

And make no mistake about it, when mid-plate earthquakes occur, they can be big. The New Madrid quake, for example, may have been a thousand times more powerful than the one in San Francisco in 1989. According to contemporary records, the shock wrecked boats in East Coast harbors and caused buildings to shake and church bells to ring as far away as New York and Boston. It even collapsed scaffolding erected around the Capitol Building in Washington, D.C. I shudder to think about what an earthquake like that would do today!

Mid-plate earthquakes don't take place just in North America, of course — they are a worldwide phenomenon. The fact that they are widespread gives scientists a chance at understanding them. By examining locations where these events have occurred, they have come to understand that they happen in places where the continental structure has been weakened at some time in the past.

The New Madrid area, for example, constitutes what geologists call a "failed rift." A rift occurs when the forces that drive the plates apart begin to operate underneath a continent. The continent is literally torn apart, and ocean moves in to the breach. The Great Rift Valley, which extends from East Africa to Jordan and includes the Red Sea, is an example of this process going on today — millions of years from now, the Horn of Africa will be an island.

A failed rift, as you might guess, is a place where this process starts but for some reason doesn't go through to completion. The result is a weakened spot on the earth's crust — a spot that is likely to give way when strain builds up in continental rocks. If you have ever watched someone cutting window glass, you can get a sense of how this works. A partial cut through the glass weakens it slightly along the line of the cut, and the glass breaks there when it is tapped. Problems remain with this explanation of mid-plate earthquakes (events in Australia, for example, seem to be different from those elsewhere), but my guess is that it is generally correct.

Geologists can measure the strain that accumulates in rocks — for example, by drilling a borehole and watching the hole deform over time. Thus, if you know where to look, you can monitor the buildup of strain. The problem is that no one knows how to use this information to predict an earthquake. For one thing, there is seldom any sign at the surface that a quiet fault exists deep underground. For another, the weakened spots we have identified, like the New Madrid fault, do not let go often enough to let us understand what triggers them. In fact, as far as we can tell, none of the big mid-plate earthquakes that have been catalogued worldwide are repeats of earlier events. So for the moment, all we can say is that quakes like the one in New Madrid could happen again — and we hope that they don't.

How Well Can We Predict the Weather?

I DON'T KNOW if you've noticed, but weather forecasting seems to have improved a lot recently. Big storms that used to blow in unexpectedly and tie up large areas of the country are now routinely predicted days in advance. Even the daily forecasts seem to be getting more accurate — how many times have you been caught by unpredicted rain during the last few years?

Historically, weather forecasting has been recognized as at best an imprecise art. It didn't really begin until the mid-1800s, when the telegraph made it possible to assemble weather information from a large area. Until the 1950s, predictions were made entirely on the basis of experience and history. There were large books of records of weather sequences, and the forecaster would look in the books for a situation like the one he saw on the weather maps, then predict that whatever happened before would happen again.

A lot of weather folklore is based on this sort of logic. When I was keeping bees in central Virginia, I had my own scheme — I noted the date in fall when the bees threw the drones out of the hives. The earlier they were expelled, the harder the winter was going to be. Nothing, it seems, is too silly to serve as a weather predictor!

In 1950, for the first time, computers capable of solving the complex equations that govern the motions of the atmosphere became available. Instead of just hoping the weather would behave as it had in the past, scientists could use the laws of physics and chemistry to predict what would happen. The last half of this century has seen a steady advance in the scope and accuracy of computer predictions.

There are, however, two basic problems with computer forecasting. One is that until recently there were never enough data to serve as input for the programs. Temperatures and wind velocities were known at only a few widely scattered points (airports, for example), which didn't provide enough data for the computers to make accurate predictions. Today, with a widespread network of ground stations and satellite observatories, we have much better data, at least in the industrialized world.

The second problem is more fundamental and appears to be built into the atmospheric system. I refer to the fact that the motions of the atmosphere may be chaotic. This is best visualized in terms of the famous "butterfly effect," so named because (in principle, at least) the atmosphere is so sensitive that a butterfly flapping its wings in Calcutta could start a chain of events in motion that might eventually cause a rainstorm in Rio.

In practical terms, the chaotic behavior of the atmosphere means it is very difficult to make long-range weather forecasts. If, for example, we make a seven-day forecast based on weather patterns at 6 P.M., then another based on patterns at 9 P.M., the two predicted patterns may be different because we started from slightly different points. This isn't a problem that can be solved with better computer programs — it's built into the earth's atmosphere. It is the reason that until 1995 the National Weather Service was willing to make detailed predictions only three to five days into the future.

In 1996, however, the NWS started to push these forecasts to seven days, with some hope of eventually going beyond that. The new forecasts are based, appropriately enough, on a new technique. Instead of making one superaccurate computer run with the best data available, scientists run the same program many times, varying the starting time or making small variations in the initial conditions. Instead of making predictions based on either 6 P.M. or 9 P.M., for example, they will run both predictions. In all, the NWS procedure calls for forty-six different computer runs. If all of the runs predict rain (to take one example), the assumption is that it *will* rain. The idea is that those features of the weather that appear in most or all of the runs are properties of the weather system itself rather than of the starting conditions for any single run. Called "ensemble averaging," this technique represents a sensible way to go about extracting useful information from chaotic systems.

A similar effort is being made for long-range climate forecasts, which the NWS is now making at least fifteen months in advance. Here the averaging takes place not over the same program run many times with different starting points, but on three quite different programs, each of which claims to describe some aspect of long-term climate. Once again, if all three approaches make the same prediction, that prediction is assumed to arise from the underlying dynamics of the atmosphere rather than from details of the means of analysis.

So over the next few years, we can expect to find out if the NWS has finally gotten better than my bus.

Was Mars Once Wetter and Warmer?

 FOR SOME REASON, the next planet out from the sun — Mars — has always held a special fascination for us. From the old obsession with "canals" on its surface to chronicling by early science fiction writers of the details of Martian civilization, people have always had some hope that the Red Planet would turn out to harbor life, as the earth does.

Alas, modern astronomy and the space program have not been kind to our dreams of Mars. The canals turned out to be optical illusions induced by the use of small telescopes, and the *Viking* landers of 1976 failed to turn up any chemical fingerprint of living things. There is indeed water on the Martian surface, mainly in the form of ice in the lower layers of the polar caps (the upper layers in winter are frozen carbon dioxide, or dry ice), but the ground temperature almost never gets above the point where water will melt. The atmosphere is thin — about what you'd have on earth at an altitude of twenty miles. Mars is not, then, a place that you would expect to be habitable today. The great question is whether it was different in the past.

Astronomers are currently engaged in a rather intriguing project — an attempt to do long-term "geology" on a planet no human being has ever visited and from which no samples have ever been taken. It may surprise you to learn that we can find out a great deal about the ancient history of Mars just by looking at photographs.

One important technique involves counting craters. Like the earth, Mars is constantly bombarded by meteors, but because of its thin atmosphere fewer of them burn up on the way down. If a surface feature on Mars (a lava flow, for example) has a high density of craters, it is reasonable to suppose that it's been there for a long time. A feature with few craters, on the other hand, would probably be relatively new.

The feature that gives most evidence that Mars was once quite different from what it is today are channels, some up to 1,000 miles long, that were almost certainly cut by running water. They resemble, in fact, the arroyos of the American Southwest — valleys cut by occasional rainstorms but dry most of the time. (Note that these channels are too small to be seen by earth-based telescopes and are *not* the legendary canals of Mars.) The valleys are clear evidence that at one time there was running water on the Martian surface, even if there isn't now. The real question is where that water came from and when it stopped flowing.

Again, satellite photographs shed some light on these questions. By looking at the shadows cast by the walls of the channels, astronomers can see that many of the channels have a U-shaped cross section, rather than the V-shaped cross section characteristic of running water on earth. In addition, they can compare the overall distribution of the channels on Mars with the drainage systems on our planet. Here, where streams are fed by rainfall, there is a pattern of small rivulets feeding into streams, streams feeding into tributaries, and tributaries feeding into large rivers. On Mars, some regions do look like this, but most of the channels appear to be fed by their own springs and to run independently of the others. This pattern has led scientists to suggest that many of the channels are "sapping valleys" — valleys fed by water seeping out of the ground rather than by rain.

The theory is that when Mars formed 4.5 billion years ago, it went through a period of being heated up, perhaps even melted, by the influx of meteorites. After the planet cooled and the surface froze, the hot magma under the surface melted ice underground. In this scenario the sapping valleys we now see on the surface were created by water that seeped to the surface and flowed for long distances.

The main problem is that according to current climate models, the air temperatures on early Mars were well below zero, so water flowing on the surface would freeze in a matter of hours. How to reconcile these results with the evidence of flowing water remains the great question about the evolution of Mars. Here are some possible solutions:

- If early Mars was really cold, perhaps water in the channels flowed under layers of ice, which insulated the top while geothermal heat warmed the bottom. Lakes with this structure are a common feature in Antarctica.

- Perhaps the climate models have left something out and there was more of a greenhouse effect than we think.

- Perhaps the presence of water produced a feedback effect, warming the climate to a point above freezing for a period of time. Proponents of this view argue that there were several outflows connected with volcanic episodes, each producing a brief "summer" and Mediterranean-sized oceans on the surface.

Is There an Asteroid Out There with Our Name on It?

JUST IN CASE you have nothing else to worry about, let me tell you about what scientists think happened on a quiet day some 65 million years ago. While dinosaurs were happily munching palm fronds and our own mammalian ancestors were scurrying around the primeval forests, an asteroid about six miles across came hurtling toward the earth. Its mass and velocity gave it the energy equivalent of about 10,000 times the entire human nuclear arsenal, and it was moving so fast that it burned a hole in the atmosphere and the ocean near Yucatán, hitting the ocean floor with scarcely any loss of speed. It kept going, plowing through solid rock to a depth of several miles, until finally all of its enormous store of energy was converted into heat.

The rock exploded, excavating a crater a hundred miles across. Some of the rock and dust blew back into space through the hole in the atmosphere before the surrounding air had a chance to rush in. This material spread out and started to fall back in, darkening the surface of the earth as it absorbed sunlight and producing an artificial night for three months. Huge tidal waves swept across the oceans. When nitrogen in the atmosphere over Central America, heated to a temperature of thousands of degrees, combined with oxygen, rain that had the corrosive power of battery acid fell over an area the size of a continent. The impact may even have triggered volcanic eruptions in India. The darkness, the rain, and the volcanoes combined to wipe out the dinosaurs, along with about two-thirds of every other kind of life on the planet.

This story is the now familiar tale of the asteroid hypothesis, in which the massive extinction of life 65 million years ago is explained by the impact of an extraterrestrial body. It is interesting in and of itself, of course, but it raises an important question: how likely is it that such a sequence of events will happen again sometime soon?

Some background: space is full of debris left over from the formation of the solar system, and the stuff is always hitting the earth. Every time you see a shooting star, you are witnessing the collision between a piece of debris the size of a pea and our planet's atmosphere. Such small bits of debris burn up completely in the atmosphere, but larger bodies burn only partially as they fall and survive as meteorites at the earth's surface. Between 15,000 and 40,000 years ago, for example, a meteorite about seventy-five feet across hit the ground near what is now Winslow, Arizona, and excavated a crater almost a mile across and six hundred feet deep.

Astronomers know that there are asteroids whose orbits cross that of the earth, which means that large impacts could indeed occur again. In 1991 an asteroid passed between the earth and the moon — a near miss by astronomical standards. And although this one was small (only about thirty feet across), the event reminds us that larger objects could collide with the earth.

This threat to our planet has been recognized only rather recently, and scientists are not exactly falling over themselves to deal with it. The first step is obvious — we need to find out how many rocks are in orbits that could endanger us. Asteroids near the earth are relatively small and thus hard to see, so this job isn't as easy as it sounds. At the moment, a few old telescopes have been converted for use in the cataloguing job. (One of those, on Mauna Kea in Hawaii, detected the near miss in 1991.) The main proposal, called Spaceguard, calls for a twenty-five-year, $50 million study to identify 90 percent of the near-earth asteroids a half mile wide or bigger. There has even been some thought about what to do if a collision becomes imminent. These schemes include blasting the asteroid with nuclear weapons, shredding it by placing a mesh with tungsten projectiles in its path, and diverting small asteroids to collide with big ones.

How seriously you take the asteroid threat depends on how often you think they're going to hit. Current estimates are that asteroids 300 feet across will hit us every few centuries (by the time such asteroids get through the atmosphere, they will be smaller than the one that landed in Arizona). Collisions with objects a mile across will happen about once every million years. Put another way, there is a 1-in-10,000 chance of such an event in the next century. If an asteroid that big hit, particularly in a populated area, it would surely cause major damage and loss of life. But we are all familiar with natural disasters (earthquakes, volcanoes, hurricanes) that occur much more often.

Catastrophic impacts, like the one that wiped out the dinosaurs, occur much less frequently, perhaps every 26 million years or so. There are a few objects as big as the dinosaur-killer in near-earth orbit, but a collision with a new comet falling in toward the sun or an asteroid newly jostled out of the asteroid belt between Mars and Jupiter is more likely. In this case, we probably wouldn't see the object before it arrived, even if we were looking for it. We'll never know what hit us!

How Will We Explore the Solar System?

THE GREAT MOMENTS in the exploration of the solar system stand out as monuments to Western civilization — the *Apollo* landing on the moon, the *Viking* landing on Mars, the *Voyager* flybys of the outer planets. But that was then; this is now. In the current political climate of the industrial world, there are real questions about whether this sort of exploration will continue and, if it does, what form it will take.

There is no danger of a total retreat from space, for there is a lot of money to be made from exploiting near-earth orbits. Communications satellites, the Global Positioning System, and "look-down" devices that monitor everything from weather to sea ice to vegetation have come to play such an important role in modern economies that they cannot be abandoned. The problem comes when we consider missions whose payoffs will be primarily aesthetic and intellectual rather than immediately practical.

The aspect of this debate that has gotten the most attention concerns manned versus unmanned spaceflight (if anyone can think of a substitute for "manned" that describes today's mixed-sex crews, please let me know). During the 1970s and 1980s, for example, many unmanned exploration missions were cut back to allow the development of the space shuttle. Why, scientists ask, should we spend billions to provide a safe environment for delicate humans in space, when machines can do the job just as well? Advocates of manned flight counter that no machine now or in the foreseeable future can match the abilities of a trained human to deal with unexpected situations. They point to the need for astronauts to repair the Hubble Space Telescope and suggest that some notable problems of unmanned programs (for example, the failure of the antenna of the *Galileo* probe to deploy) could have been fixed easily had there been a crew on board.

But I think that this debate, rooted as it is in the experience of past decades, has pretty much played itself out. All participants now recognize that both manned and unmanned flights have a role to play, and there may even be a consensus developing about reasonable levels of funding for each type.

But as one great debate winds down, another has begun. Past space missions have been pretty expensive. The *Voyager* probes, launched in 1977 and now heading out of the solar system, have cost over $2 billion so far. The *Galileo* mission to Jupiter cost over $3 billion. The question: what can we do in the future, when funds for these sorts of missions just aren't available?

There are many reasons why missions like the *Galileo* cost so much. As NASA budgets have been cut over the past years, the number of launches has dwindled. This means that each launch becomes the last bus out of town and everyone piles on. The result: the complexity and cost of each vehicle grows, and when the inevitable accident happens (as it did when the *Mars Observer* disappeared in 1993), the cost is measured in billions of dollars.

The new trend is toward spacecraft that are, in the current NASA buzzwords, "smaller, faster, cheaper, better." Called the Discovery program, this new generation of spacecraft will hark back to the early days of space exploration, when small, single-purpose satellites were launched regularly. Of course, "cheap" is a relative term. The probes NASA has in mind will cost no more that $150 million, which isn't exactly chicken feed. They will be much smaller than current probes. *Galileo,* for example, weighs about three tons, but the *Mars Pathfinder*, to be launched at the end of 1996, weighs only about nine hundred pounds (not counting fuel). The idea is that many smaller missions can be run for the cost of one large one, and even if a few of the smaller missions fail, the net result is more "bang for the buck" with less risk of catastrophic loss.

The prototype for a scaled-down space program was the launch of a probe called *Clementine* in 1994. It was designed primarily to demonstrate the capability of a new generation of lightweight sensors that were developed in the Star Wars program, but its scientific mission was first to map the moon, after which it would fly by a near-earth asteroid. At a cost of $80 million (a figure that doesn't include developing the instruments), *Clementine* fits well into the Discovery program's limits. It did, in fact, produce the first digital map of the moon, along with the first map of rock types over the lunar surface and the first detailed maps of the polar regions. Unfortunately, an error by an onboard computer caused the probe's jets to expend all their fuel, and the second part of the mission had to be abandoned. But we had already gotten our money's worth from *Clementine*.

So in the future, you can expect to see small missions to Mercury, Pluto, and Venus, and many missions to study nearby asteroids and comets, missions that may even bring samples back. And all of this will be done by a new generation of cheap, "throwaway" spacecraft.

What is the Earth Observing System?

ONE STANDARD criticism of the space program is that it diverts scarce brainpower and financial resources from pressing problems on our own planet. I find this criticism somewhat misguided because some of the most important consequences of the space program have been discoveries that increased our understanding of the earth. We want (and need!) to answer questions about the way our planet works, and often the questions are best answered in space. To understand what I mean, you have to know something about how we have acquired knowledge about the earth and its systems in the past.

Suppose you want to know the temperature of the ocean surface. Up until the late 1970s, the only way you could get such information was to look at measurements taken by ships at sea, and there are problems with these data. First, there are almost no data about areas far from shipping lanes, such as the Central Pacific or the South Atlantic. Second, except in regions of intense traffic, the data don't tell you how the temperature varies over time. Satellite data do not suffer from these sorts of drawbacks, for satellites cover the entire planet and take continuous measurements over long periods of time.

In the late 1970s we began launching satellites whose job was to look back at the earth. A series of satellites in the 1980s developed the first record of sea-surface temperatures that took in the entire planet. Other satellites, taking advantage of the "whole earth" picture that can be seen only from space, produced records of phenomena such as ocean wave height, abundance of phytoplankton, and patterns of vegetation and ice on land. It was these early satellites that indicated the southward expansion of the Sahara Desert. They also provided the crucial ongoing data needed to monitor the Antarctic ozone hole and to establish its cause.

The original plan called for NASA to launch the first satellite in what will be called the Earth Observing System (EOS) in 1998. Called *AM-1,* this satellite is supposed to be a platform with five different instruments that will continuously monitor the solid earth, the oceans, and the atmosphere and supply us with worldwide data on the aerosol content of the air, the humidity, and other important features of the earth's systems. Political pressures related to the high cost of such systems, however, has put the future of this plan in doubt, and NASA is now looking at the possibility of launching a series of smaller, cheaper satellites to gather the same sorts of data.

The Earth Observing System is part of what NASA likes to call "Mission to Planet Earth." The purpose of the satellites is to establish, for the first time, a global base line of information about quantities that are important in understanding how the earth works. One of the most difficult questions we face has to do with whether or not our climate is changing, and if it is, whether that change is due to human activities. We are handicapped at the moment because we simply do not know what the base line is — how can you tell if something has changed when you don't know what it was to begin with?

There are, of course, many technical difficulties involved with putting a satellite system like this in orbit. Let me tell you about one problem that people really hadn't thought about before the EOS became a reality, and that is the prodigious rate at which information will arrive from the satellites. When the full suite is in orbit, the data coming down will have to be processed by earth-bound computers at a rate equivalent to three sets of *Encyclopaedia Britannica* per minute. One of the great challenges faced by NASA and the scientific community is to develop the technological know-how to deal with data at that rate. (Picking meaningful information from such a stream is a little like trying to get a drink from a fire hose). Important problems include representing the data (probably in pictures) in a way that makes sense to human observers (nobody wants to dig through mountains of computer paper to find the numbers he or she needs); developing "smart" instruments on the satellites that will know which data are important and should be sent down to earth to be processed and stored, and which data can be thrown away at the beginning. In addition, we need to train a corps of scientists who know enough about the origin and importance of the data to know when a particular reading should be flagged. All these problems are being worked on, and I expect that by the turn of the century they will be solved.

And, as often happens in these sorts of situations, when scientists learn how to deal with information of this volume, the technology will become available to the rest of society. Because of the EOS, our coming trip down the information superhighway may be a little less bumpy.

So What about That Face on Mars?

ON JULY 26, 1976, the spacecraft *Viking I* passed over a desert region in Mars's northern hemisphere — a region called Cydonia — and snapped a picture of the surface. When NASA scientists developed the picture, they saw something that sort of resembled a human face, and they showed the photo to reporters for a laugh. Thus was born the Face on Mars — one of those stories that lead a shadowy existence on the fringes between science and pseudoscience, occasionally popping up into public attention for a short time, only to be forgotten again.

Now don't get me wrong — I'm not suggesting that the question of whether some ancient civilization carved a human face (!) on Mars ranks as one of the burning scientific questions of the age. It doesn't. I include it here as a proxy for all of those issues — the Loch Ness Monster, UFO abductions, ESP, and, until recently, the Shroud of Turin — that are rejected and ignored by almost all scientists but that nevertheless won't go away.

At one point in my life I devoted a fair amount of time to investigating a few questions of this type (ancient astronauts and pyramid power, if you must know). Because of this experience, I tend to be fairly skeptical of the claims about such phenomena: upon close examination most of them turned out to be a lot less defensible than they seemed at first. To my mind, the Face on Mars fits this pattern perfectly.

What are the facts? The primary purpose of the *Viking I* photographic survey of the Martian surface was to help scientists choose a landing site for the *Viking* lander that was already on its way to the Red Planet. Regions like Cydonia were too hilly and broken up to be of much interest for a landing site. So when a blurry photograph showed a mesa about 1,500 feet high and a mile long, with what looked like a human face on it, the NASA people didn't pay much attention.

As usually happens in this shadow world, a few people looked at the photographs and decided that vital information was being covered up. As far as I can see, four types of arguments are being used to buttress the claim that there is more to this issue than mainstream science is willing to admit. These are (1) use of an image-processing technique (called "shape from form") to show that the face is not the result of a trick of the light but corresponds to real formations on the mesa; (2) analysis of nearby formations that purports to show many unusual structures (mainly pyramids) in the area; (3) an argument that angles drawn between some neighboring structures repeat themselves; and (4) a study based on a mathematical technique using fractals, which claims that the structures don't have the types of patterns typical of the area around them.

When I was investigating the pyramids, the first thing I did was to find out whether the claims made about them were in fact true. Locating a survey of Giza, I quickly found that (1) the pyramids are not laid out on perfect squares, and (2) their sides do not point exactly north-south. These findings demolished two of the most common claims made by believers in pyramid power. If I were going to take the Face on Mars seriously, I would redo all of the analyses to make sure they were done right. I don't have time to do this, so let me grant that all claims are true. Do they show a reasonable probability that there was an advanced civilization on the Martian surface?

In a word, no. The shape-from-form analysis, for example, just shows a certain placement of features on the mesa and does not indicate that they are artificial. Indeed, the laws of probability dictate that any planet must have some chance arrangements of geological features that humans will perceive as pictures. Are pyramid-shaped structures indicative of conscious construction? Geologists know of many such formations in places on earth where the primary erosional force is the wind (think of the Badlands of South Dakota). The repeating angles? This argument might have some force if these structures were out on an open, featureless plain. In fact, they are in an area where there are many fissures and therefore many angles. In this kind of situation, some will be numerically similar just by chance. And as far as the study showing that the mesa is different from the surrounding area, what if it is? That just proves it's different, not that it was built by conscious design. I can think of plenty of places in the American West that would meet this criterion.

I expect the Face on Mars to hit the news again when NASA resumes missions to Mars in 1998. As with *Viking*, there will be some photographic surveying and, I suspect, loud calls from the fringe to get a better picture of Cydonia. If it's possible to do this, it may be done, but with the new lean, cheap approach to space exploration, it may not be. And if those probes don't come back with pictures of this mesa, the cries of "coverup" will start all over again.

5

Biology
(Mostly Molecular)

How Are Genes Controlled?

 HERE'S ONE of the most baffling riddles in biology: every cell in an organism, for example a human being, contains the same DNA. Cells in your brain and intestines all contain the gene for making insulin, but none of these cells actually makes insulin — that task is reserved for cells in your pancreas. So how do cells know which genes are supposed to be operating? How do cells in the pancreas know that they're supposed to activate the gene for insulin, while cells in the brain know they're not supposed to? This is the problem of gene regulation.

The first explanation of gene regulation dates back to the 1950s. It involved genes in bacteria, where the DNA floats loose in the cell instead of being contained in a nucleus, as it is in advanced organisms. The bacterium studied was *Escherichia coli*, which resides in the human digestive tract. Instead of having to decide whether it's in the appropriate place to make a particular protein, as a cell in the human body does, the bacterium cell has to decide when to make a protein.

E. coli, for example, has a couple of genes that allow it to make a protein to break down lactose, a sugar found in milk. But milk is not always present in the human digestive tract — it tends to come in spurts separated by long periods when it is not present. The bacterium, then, has to have a way of switching on the lactose-digesting genes when the milk is there and turning them off again when the milk is gone.

Every gene on bacterial DNA consists of a long string of molecules that code for the lactose-digesting protein, preceded by a short stretch of molecules called a "promoter." This is where the enzymes involved in assembling this particular protein attach to the DNA and start moving down the double helix. Think of the promoter as being analogous to the ignition switch on your car — it has to do its job before anything else can happen. Between the promoter and the gene proper is another short region called an "operator." Under normal circumstances, the enzymes pass over this region on their way to the gene.

When no lactose is around, however, a "repressor" molecule wraps itself around the operator region, blocking the advance of the enzymes and preventing the genes that follow it from being read. Think of it as a clamp holding the two strands of the double helix together and blocking downstream traffic. So long as it's in place, the proteins that break up lactose can't be made. If you drink some milk, however, the lactose molecules bind to a specific spot on the repressor molecule, causing it to flex and lose its grip on the DNA. The repressor-plus-lactose floats off, and the cell begins to churn out the proteins to break down the lactose. When the lactose has been digested, the repressor (now minus its lactose molecule) floats back and reattaches, shutting the whole operation down until your next milkshake.

The mechanisms of gene regulation in simple bacteria have been pretty well worked out, and all have the kind of direct, relatively uncomplicated nature as that of lactose metabolism in *E. coli*. When it comes to the more complex cells in human beings, however, the mechanisms begin to get very strange. Our genes don't have the simple promoter-operator-coding sequence described above. Instead, it seems that in addition to the promoter, other regions of the DNA molecule located far upstream from the main part of the gene can affect whether that gene is activated. These regions are called "enhancers" and "silencers." Think of them as analogous to the gas pedal and brakes on your car. At all of these sites a large family of molecules can attach themselves and affect what happens next, and control seems to be exerted by a complex (and not fully understood) series of "start" and "stop" commands. Here's a list of the relevant molecules:

- *activators* — these bind to enhancer sites and determine which genes are switched on.
- *coactivators* — these bind to activators (which are bound to enhancers) and link them to
- *basal factors* — a complex set of molecules that bind to the promoter region of the DNA, providing a link between activators, coactivators, and the DNA itself. They seem to get the enzymes that will read the DNA into position and send them on their way.
- *repressors* — these link to silencer regions to shut down (or slow down) genes. They may also link to coactivators.

So each gene in your cells has this complex of molecules that hover around it, and the actual molecules in the complex are different for each gene. Over the last few years scientists have started to see the broad outline of how these four types of molecules link together to turn genes on and off. We've come that far along the road to answering the question of how genes are regulated. The next question — and in some ways the more interesting one — is how cells know which activators and repressors to make. Each of these molecules is coded for by a gene somewhere in the DNA, and sorting out how cells know which of *those* genes to turn on is, I think, where the action is going to be in the coming years.

How Does an Organism Develop
from a Single Fertilized Egg?

ALL OF US started as a single fertilized cell in our mother's Fallopian tubes. Over a period of nine months or so, that cell grew, through a process of repeated cell division, into a fully formed newborn. The same kind of development is seen in every multicelled organism on the planet. The question of how this process takes place — how the fertilized egg knows what it's supposed to do and how it does that — is the domain of developmental biology. Like everything else connected with the life sciences, this field is going through a period of explosive growth as people learn how to relate biological processes to the underlying molecular interactions.

The general pattern of development has been known for some time: first, the initial cell undergoes a few divisions, then the ball-shaped embryo starts to fold and deform in a process known as gastrulation, and, finally, organ systems begin to form in the process called morphogenesis ("creation of forms"). When these processes are examined at the molecular level, two very interesting truths emerge. (1) Like a sculptor who uses just one set of tools to form all of the different parts of a statue, nature seems to use the same proteins (which you can think of as molecular hammers and chisels) to perform many different functions as the organism develops, and (2) the same basic processes seem to take place in many species, with similar genes coding for similar functions in organisms as different as mice and fruit flies.

Let me tell you about some recent work in this field to give you a sense of what is being done. One of the basic problems facing the fertilized egg is making sure that the adult organism will have a well-defined body plan. There must be a top (head) and a bottom (feet), a front and a back. How does a single spherical cell, floating in fluid, know which direction is "up" and which "down," which direction is "front" and which "back"? For one organism, the fruit fly, we actually know the answer to these questions.

Before the egg is fertilized, the female fruit fly deposits some RNA molecules in it. One type of RNA is concentrated at the end of the egg that will eventually be the head, another kind of RNA at the other end. After fertilization, the RNA molecules at the head produce one specific protein in its part of the embryo, while the RNA molecules at the feet produce another. The proteins at the head turn on a particular gene in the fruit fly DNA, while those at the feet shut that gene down. Thus the top-bottom axis is established in the embryo, a process that eventually leads to the establishment of the three segments characteristic of the insect body. One goal of developmental biology is to produce a similarly detailed account of how this process goes on in other organisms, including humans.

Questions about how organs form are also being addressed at the molecular level. For example, how do cells in the embryo communicate with each other? Classic experiments in developmental biology long ago established that certain cells, if transplanted from one part of an embryo to another, could "recruit" cells in their new home into forming organs appropriate to the original site. The embryo might grow a second body axis, for example. Obviously, the transplanted cells communicated something to their new neighbors. The question is what was communicated and how.

We know that cells in the embryo live in a rich biochemical environment, with molecules being secreted by some cells and taken in by others. Scientists are now beginning to trace out the details of how these chemical pathways operate — how a cell recognizes a molecule and takes it in, how the signal carried by the molecule is transmitted to the cell's DNA, how the signal turns on specific genes, and what the effects of those genes are on the developing organism. The signaling process is rarely simple, for each cell is receiving many signals at the same time, and the final outcome is generally the result of a complicated interplay between them. There are, for example, no fewer than four molecules that, by alternate attraction and repulsion, guide the way nerve cells connect to each other.

The first practical effect of our new understanding of the chemical basis for embryonic development will probably be drugs that have the ability to turn on genes in adult cells. Cells in the embryo have powers of growth not present in the adult — if a fetus *in utero* suffers a cut, for example, it will heal without a scar. If we know which molecules trigger cell proliferation in the embryo, it might be possible to apply those molecules to adult cells to bring about the same effect. There are already a few drugs like this in use, such as drugs that stimulate the production of red blood cells to combat anemia and white blood cells after bone marrow transplants. These are just the beginning, though. Scientists speak confidently of having drugs that will stimulate the repair of the retina of the eye after certain kinds of damage; promote the healing of skin ulcers, which are a common side effect of diabetes; promote the healing of difficult bone fractures; and, perhaps, allow nerve cells to grow again after injuries to the spinal cord. All of these drugs are just possibilities at the moment, but I would be amazed if at least some of them didn't become realities in the next few years.

How Is DNA Repaired?

EVERY CHEMICAL reaction that goes on in every cell in your body is programmed by your DNA. The instructions contained in the sequence of molecules that make up the double helix tell your cells when to make certain chemicals, when to divide, and when to stop dividing. Damage to DNA is implicated in the onset of diseases like cancer as well as in ordinary aging. Maintaining the integrity of the messages written in DNA, then, is a crucial part of the functioning of every living thing.

Unfortunately, the importance of DNA has given rise to some folklore about the molecule's fragility. DNA is pictured as a passive victim of environmental forces, any of which can destroy the cell and, perhaps, the organism of which it is part. But as a moment's reflection on the nature of evolution shows, importance and vulnerability are not the same thing. Organisms whose cells develop means of repairing damage to their DNA have an obvious survival advantage over those that do not, so it should come as no surprise to learn that human cells are able to make such repairs.

What is surprising, however, is the extent to which DNA damage occurs in the normal course of affairs. In the late 1980s, scientists using extremely sensitive chemical tests concluded that each cell in your body, on the average, sustains about 10,000 "hits" to its DNA each day. Some of this damage comes from influences in the outside environment, but most is chemical damage done by the byproducts of ordinary cellular metabolism. One major culprit, for example, is the class of chemicals known as oxidants, which are produced in the normal "burning" of carbohydrates to produce energy in the cell. (Broccoli and related vegetables contain antioxidants, which is one reason they figure prominently in anticancer diets.) Normal cells, then, live on a knife edge in which damage is constantly occurring and constantly being repaired.

Scientists have observed two broad types of DNA damage, each of which has its own repair mechanisms. One of these, "mismatch repair," deals primarily with mistakes that occur when DNA is being copied during cell division. If you think of the DNA molecule as a twisted ladder, the molecule copies itself by cutting through the rungs, then having each half reassemble its missing partner from materials in the cell. One type of mismatch occurs when one half of the ladder slips in relation to the other during reassembly, leaving a loop of unmatched half-rungs sticking out to the side of the DNA. Another type of mismatch involves building the new half-rung from the wrong kind of material.

In mismatch repair, it seems that some half-dozen genes produce proteins that recognize the mistake and cut out the appropriate stretch along the side of the double helix. A small number of other proteins supervise the reconstruction of the missing segment. Without these proteins, miscopied DNA cannot be corrected, and mistakes, once made, will be passed on to future generations of cells. A defect in one of the genes that produces repair proteins is now known to be a cause of a very common kind of colon cancer in humans.

A more complex kind of repair process, "nucleotide excision repair," deals with damage to DNA caused by agents such as chemicals from both inside and outside the body and radiation, including ultraviolet radiation in sunlight. Think of it as a "jack-of-all-trades" that will fix whatever kind of damage it encounters. Its basic operation is similar to that of mismatch repair — a defect in DNA is detected, the "rungs" of the DNA ladder are opened up, the side with the defect is cut out, and, finally, a nondamaged strand is reassembled to "heal" the molecule. This process is used to effect a wide variety of repairs — researchers have yet to find damage to DNA that can't be repaired, other than a complete breaking of the strand. But other than the fact that it exists, we know little about the detailed mechanism of nucleotide excision repair.

Our understanding of how cells repair DNA is going to increase rapidly in the coming years. In the short term, this will have effects in two areas. In medicine, more diseases caused by failures in the mechanism will be identified, and diagnostic tests (and perhaps even treatments) will become part of your doctor's art. In public policy, an understanding of DNA repair will have a major impact on environmental debates. Up to now, whenever we have tried to assess the risks of exposure to chemicals or radiation, we have assumed that there was no minimum safe dose — that even a tiny exposure could cause damage to cells. In the absence of information about how cells deal with damage, this is the model a prudent person has to follow, but it surely overestimates actual risks. An understanding of complex DNA repair mechanisms will allow us to determine the level of exposure below which ordinary repair mechanisms can deal with damage. Folding that information into our risk assessments could result in a considerable loosening of regulatory standards in many areas.

Why Do Cells Die?

 DURING THE fifth week of a pregnancy, the hands of the human fetus are flat disks shaped something like Ping-Pong paddles. Over the next few weeks, cells in what will be the spaces between the fingers die, and the hand continues to grow in its familiar five-fingered form. This is an example of one of the most remarkable processes in living systems — the programmed death of cells. Another example: a tadpole's tail disappears as the tadpole grows into an adult frog because millions of cells in the tail are programmed to die.

Programmed cell death is called "apoptosis" by scientists (the second "p" is silent). The word comes from the Greek *apo* (away from) and *ptosis* (falling). It is different from accidental or pathological cell death, which results from disease or injury. In apoptosis, the cell breaks up in an orderly way and its materials are absorbed by healthy neighbors.

Over the past few years, scientists have come to realize that every cell is capable of undergoing apoptosis — in effect, each carries a "suicide gene" that, when activated, kills the cell. Some scientists go so far as to talk about this process as a necessary aspect of the "social contract" involved in having so many cells living together in a single organism; a cell's self-sacrifice is often necessary to the overall development of the organism. In the development of the human nervous system, tendrils on nerve cells grow out along the track of molecules emitted by specific sensory cells — for example, nerve cells in the brain connect to the retina of the eye by this process. If a cell can't find the appropriate place to hook up or if it gets into the wrong region of the body, it commits suicide, while those that make the proper connection survive.

Early work on the process of cell death used an interesting model, a microscopic worm called *Caenorhabditis elegans,* or *C. elegans* for short, which is a laboratory favorite of biologists. All the adult worms have exactly 1,090 somatic cells, and during their maturation exactly 131 of these undergo apoptosis. Using this astonishingly precise example, scientists were able to identify the protein molecule that initiated cell death and the gene that coded for the protein. Then, in one of those fascinating scientific developments that is like a good detective story, different groups of scientists located similar genes in other kinds of organisms (including mammals) and found an amazing evolutionary overlap — suicide genes seemed to be very similar in widely different organisms. In one classic experiment, a mammalian gene was substituted for the original gene in the DNA of *C. elegans*, without producing any change in the cell.

There are two fundamental questions we can ask about apoptosis: (1) what kind of signal does the cell have to receive to initiate the process, and (2) once the signal is received, what is the mechanism by which cellular suicide is carried out? Research in these areas is one of the hottest topics in the life sciences today.

As far as the first question is concerned, the earlier example of nerve cells in the brain may point to an important truth. Cells receive signals in the form of specific molecules from other cells, and a consensus seems to be developing that these signals keep the suicide genes from activating. If this notion is correct, you can think of each cell in your body as sending out molecules that tell its neighbors not to kill themselves and receiving similar signals from them. When a cell no longer gets those signals, it knows its time is up and initiates apoptosis.

The details of the internal mechanisms of the cell's signaling system are now starting to become clear. It appears that in mammals apoptosis is not just a simple matter of turning on a switch. In fact, several different molecules in our cells seem to be associated with the process (each molecule coded for in a separate gene). Some of these molecules seem to turn the cell-death mechanism on, while others seem to protect the cell. It's almost as if the cell contained a dial whose setting was determined by the properties of "death" and "antideath" materials. When the suicide signal arrives, a cell in which the death materials predominate will start the suicide process, while one in which antideath predominates will simply ignore the signal. The gene that integrates these signals was identified in 1995 and was named, appropriately enough, the "reaper" gene.

Work on apoptosis is not motivated simply by scientific curiosity — there are important medical questions as well. For example, one antideath gene identified in cells is called p53; it has long been known that the absence of p53 is associated with the onset of tumors. This suggests that some cancers may result from the failure of cells to die rather than from uncontrolled growth. Premature triggering of the suicide response may also be implicated in Parkinson's and Alzheimer's diseases. As so often happens when we investigate basic processes in nature, practical implications follow naturally.

Why Does a Protein Have the Shape It Has?

 LIFE IS BASED on the chemical reactions between molecules, and these reactions depend on
molecular geometry. This statement represents
one of the most profound truths we know about
the nature of life.

Take the combining of two large molecules as an example.
Bonds can be created only between individual atoms. Think of
the atoms that could form bonds as small patches of Velcro on
a large, convoluted sphere. For combination to occur, the molecules have to come together in such a way that the Velcro
patches are juxtaposed. The chance of this happening in a random encounter is pretty small, so chemical reactions in cells
depend on a third kind of molecule called an enzyme. The enzyme attaches separately to each of the molecules taking part
in a reaction, assuring that each one's Velcro is in the right position and allowing the reaction to proceed. The enzyme is not
itself affected in the process. Think of it as a broker who brings
together a buyer and a seller but doesn't buy or sell anything
himself.

The enzymes in our cells are proteins, which are long molecules made from a set of smaller molecules called amino acids.
The amino acids are assembled like beads on a string, and the
resulting protein then curls up into a complex shape. Because
there are so many possible combinations of "beads," the final
assembly can have many possible shapes, which makes proteins ideal for the role of enzymes.

Cells work like this: the DNA in the nucleus contains codes
specifying the order of the amino acid beads that go into a

protein. The stretch of DNA that contains the blueprint for one protein is called a gene. Each gene codes for one protein, and each protein acts as an enzyme for one chemical reaction. In a particular cell, as many as a few thousand genes may be operating at any given time.

This knowledge, combined with our ability to manufacture genes and implant them in bacteria, opens an exciting possibility. If we know the chemical reaction we wish to drive, we can figure out the shape of the enzyme needed to drive it. If we know how a specific sequence of amino acids folded up into a protein's final shape, we can design a gene to make that sequence, put it into some bacteria, and brew it up.

But there's a problem with this notion that has plagued biochemists for the last forty years. Even if we know the sequence of amino acids in a protein — the order of the beads on the string — we simply do not know how to predict the protein's final shape. A solution to what is known as the "protein folding problem" remains tantalizingly beyond the grasp of modern science.

The reason for this gap in our understanding is simple: there can be hundreds of thousands of atoms in a single protein, and even our best computers aren't good enough to keep track of everything that goes on when the protein folds up.

At the moment, two lines of research are being pursued. The first involves experiments whose goal is to designate intermediate states in the folding process. For example, a long chain might first twist into a corkscrew, then have some segments fold over double, and then fold up into its final shape. By knowing these intermediate states we can break the folding process down into a series of simpler steps. One difficulty with this approach is that proteins apparently can follow many different folding sequences to get to a given final state.

Other scientists are trying to use clever computing techniques to predict the final shape a string of amino acids will take — techniques that do not require following each atom through the folding process. For example, computer programs can estimate the final energy state of different folding patterns. Since systems in nature move to states of lowest energy, the suggestion is that when you find the lowest energy state, you have found the final pattern of the molecule. The problem: there may be many low-energy states, and it becomes difficult to know which one the molecule will wind up in.

Another computer approach involves the techniques of artificial intelligence. Data on known folding patterns of amino acid strings are fed into a computer, which then guesses a folding pattern for a new molecule based on analogies to known proteins. The problem: you can never be sure the guess is right.

Whichever technique finally brings us to a solution of the protein folding problem, one thing is clear. When the problem is solved, we will have eliminated a major roadblock on the road to manufacturing any molecule we want.

The Miracle Molecules

 EVERY JANUARY *Time* magazine names its Man or Woman of the Year. Not to be outdone, *Science* magazine, the publication of the American Association for the Advancement of Science, has introduced its own version of this award. Every December, *Science* names a "Molecule of the Year." The award was kicked off in 1993 by honoring molecule p53.

The "p" stands for protein, the "53" for the fact that the molecule weighs 53,000 times as much as a hydrogen atom. It was discovered in 1979, almost in the Dark Ages of molecular research. Although it was known to have some association with cancer, it was largely ignored until 1989, when researchers found that mutations of the gene that codes for the protein were common in colorectal cancers and that normal genes acted as tumor suppressors. Today scientists estimate that 50 percent (and perhaps as many as 80 percent) of human cancers involve p53 in some way. (A matter of nomenclature: the gene that codes for p53 goes by the same name as the molecule.)

The best way to describe what has happened in research on p53 is through an analogy. Suppose we sent four teams of searchers into a city. The first team was to locate the most important businessman, the second the most important government leader, the third the most important artist, and the fourth the most important scholar. Suppose each team pursued its independent inquiries, and at the end, they all found themselves converging on the same house on the same street. We would probably conclude that the occupant of that house was an extraordinary individual.

Something like this went on in the 1990s with p53. Researchers in four totally different fields of biology found that the gene and its protein played a crucial role in the phenomenon they were studying, and were subsequently surprised to learn that it was important in another area as well. Each of these four fields is discussed in more detail elsewhere in the book. They are:

DNA Repair. When damage occurs to DNA, p53 stops the cell from dividing until the repair is made and also marshals the enzymes that actually make the repair.

Apoptosis. If the damage to DNA is not repaired, or if the cell must die for the overall good of the organism, it is p53 that triggers "cellular suicide."

Tumor Suppressors. Cancer researchers have long known that some molecules act to suppress the growth of tumors. Given its role in DNA repair, we should not be surprised that p53 belongs to this select group.

Cell Cycle Regulation. Scientists investigating the way cells divide have found that p53 plays a crucial role in the process, particularly in shutting it down.

In the end, p53 emerges as a kind of master control molecule in the cell; its task is to ensure that the cell's descendants get the same DNA that the cell inherited. Failing this, its job is to make sure that a defective cell doesn't pass damaged DNA on to future generations.

The implications of the functioning of p53 are obvious to medical researchers, and a race is under way to see who will be the first to exploit it in the treatment of cancer. We can already identify specific changes in the p53 gene caused by carcinogens in tobacco smoke, and we can trace their effects to the development of lung cancer. The molecule's role in the development of other forms of cancer is being worked out.

One approach to treating cancer would be to use gene therapy, perhaps with manufactured viruses, to inject the healthy p53 gene into tumor cells. Experiments in the laboratory indicate that as soon as tumor cells acquire an adequate supply of p53, they commit suicide. Another approach is to flood a tumor with p53 molecules in the hope that they will be taken into the cells and produce the same result.

Finally, the existence of a molecule like p53 poses a serious problem for evolutionary theory. If this single molecule is so important, why hasn't natural selection led to the development of a backup system that could take over (at least in part) when the gene fails? Such backups exist for less crucial molecules, and it's astonishing that there isn't one for p53.

Whither the Human Genome Project?

 AT THE END of the 1980s there was a major debate in the scientific community over the Human Genome Project (HGP), whose purpose was to produce a complete reading of human DNA (the sum total of any organism's DNA is called its genome). Some scientists objected to the project on the grounds that it would be too boring to attract good minds, among other things. Others raised ethical and political issues. Despite these objections, the Human Genome Project was founded by Congress in 1990 as a fifteen-year, multi-billion-dollar enterprise. I'm happy to report that now, better than one-third of the way into the project, none of the objections have turned out to be valid. Advances in technology have allowed most of the boring work to be handed off to machines, and many of the best scientific minds in the country are involved in the HGP in one way or another.

Here's what the project is about: the DNA of every cell in every human being contains the blueprint for turning a single cell into a functioning adult. One human differs from another by only a fractional percentage of that blueprint, which means that you only have to "read" the DNA blueprint once, and then you have the information forever. Furthermore, as we are now beginning to realize, virtually every human disease has a genetic component, so being able to read the blueprint will have enormous medical payoffs.

The easiest way to picture DNA is as a long, flexible ladder, with each rung made up of two molecules (called bases) locked together. If you take that ladder and give it a little twist, it produces the familiar double-helix structure of DNA. Certain stretches of DNA — anywhere from a few thousand to a few hundred thousand bases long — are called genes. The sequence of bases on these genes contains the instructions for the production of proteins that will control the chemical reactions in the cell. In human cells, the DNA is wrapped in bundles called chromosomes. Most cells contain forty-six chromosomes: twenty-three from the mother and twenty-three from the father.

There are two ways to explore DNA. A process called "mapping" involves finding the placement of specific genes on specific chromosomes. You can think of mapping as a rough first exploration of the genome — something like the Lewis and Clark expedition to the American Northwest. Another process, called "sequencing," involves specifying the actual rung-by-rung order of base pairs in the DNA "ladder" — rather like the National Geological Survey that followed Lewis and Clark. The goal of the HGP is to produce both an exhaustive map and a complete sequencing of human DNA.

To establish that a particular kind of defect in a particular gene is responsible for a particular disease, it is necessary to (1) find which chromosome the gene is on, (2) find the exact location of the gene on that chromosome, (3) sequence the gene, and (4) identify the "misspelling" associated with the disease. For example, in 1989, the cause of the most common form of cystic fibrosis was identified as the absence of three specific base pairs on a specific gene of chromosome seven, and this knowledge has already led to experimental therapies for the disease. This is a good example of what we can expect from the HGP.

But once the Genome Project is completed, we will be faced with a new set of problems. If, for example, conditions like alcoholism and propensities for any number of diseases are genetically based, what should we do with that information? Most people (myself included) would want to know if they had a genetic propensity for a disease if they could do something about it. For example, someone with a known propensity for a particular cancer could arrange to have regular tests so that if the cancer did develop it could be caught early, with a high chance of successful outcome. On the other hand, suppose someone analyzed your genome and told you that no matter what you did, you had a fairly strong chance of contracting an incurable disease before your fiftieth birthday. Would you want to know that? I certainly wouldn't. And then there are the questions of who owns the information on the genome, an issue discussed elsewhere in this book.

In any case, the Genome Project is moving along. In 1995 the first complete genome of a living organism (a simple bacterium) was published; in 1996, the 12-million-base genome of brewer's yeast. The main points of discussion these days are, as you might expect, cost (current goals are in the range of 20–30 cents per base) and accuracy (the goal is 99.99 percent). New automated techniques capable of meeting these goals are expected to produce sequences of about 3 percent of the human genome by 1999. But the work will go steadily forward, and sometime after the turn of the century, we will have the reading of the entire human blueprint. Before that time maybe we should start thinking about what we're going to do with it.

What Are the Ethical Standards
for Molecular Knowledge?

 WHEN THE Human Genome Project is complete and the entire story written in human DNA is recorded, when we understand the molecular mechanisms causing human disease and (possibly) behavior, we will not have come to the end of our problems. We will have a totally new sort of knowledge about individual human beings, but this knowledge will open all sorts of new questions in the domain of ethics. Broadly put, our focus will have to shift from "How does it work?" to "What are we going to do with this knowledge?"

Some of these issues have already started to appear in legal debates. If you know that a particular defect in a particular gene predisposes you to a particular type of cancer, are you obligated to share that information with your insurance company and risk having the defect classified as a preexisting condition or, worse yet, being denied insurance coverage altogether? Can insurance companies or employers demand that genetic tests be done as a condition of insurability or employment? These are difficult issues, but the legal system can deal with them relatively easily. Other issues, however, go well beyond the legal ones. To test your own "genetic ethics quotient," try dealing with the following illustrative problems posed recently at a conference on genetics at the Jackson Laboratory in Bar Harbor, Maine.

Case 1

A woman is diagnosed with colon cancer and when tested is found to have a genetic defect associated with susceptibility to the disease. There is a history of colon cancer in the woman's family, which makes it likely that she inherited a mutated gene and that she may have passed it on to her children. The woman, however, does not wish to tell her family (including her brothers and sisters) about the diagnosis and refuses to allow her children to be tested for the defect.

In this case, there is a conflict between the woman's right to privacy — her ownership and control of her own genetic information — and the safety of her siblings and children. Should they be told or not?

Case 2

This is actually two cases, probing two different aspects of the same issue.

(a) A woman knows that there is mental retardation in her family and that she herself carries a recessive gene responsible for the condition. She requests genetic testing of the fetus she is carrying. Her intention is to seek an abortion if the testing shows that the child will be retarded. She also announces that she will abort an otherwise healthy female fetus if it is, like her, a carrier of the condition.

(b) A husband and a wife, both dwarfs, seek genetic counseling when she becomes pregnant. The gene responsible for their dwarfism is known. Children who inherit two flawed copies of the gene normally live only a few years, and children with one flawed copy will be dwarfs. The couple plans to abort the fetus if two flawed genes are found. They also announce that they will abort the fetus if it hasn't inherited one flawed copy of the gene — that is, if it will not be a dwarf.

In both cases genetic information will be used to abort an otherwise normal fetus. In case 2a the woman's desire is to end the history of mental retardation for her descendants. In effect, she is taking upon herself the decision that a daughter with the mutated gene would have to make about whether or not to have children. In case 2b the dwarf couple want their children to be like them, even though many such couples have raised normal-sized children.

We already know about thousands of genetic conditions, and all of us carry mutated genes of one sort or another. These facts raise very difficult questions in the area of abortion policy. Most people who do not oppose abortion on moral grounds would probably feel that aborting a fetus with a serious genetic defect like mental retardation is justified, but what if, as in case 2, the fetus carries a single copy of a defective gene? How about abortion decisions based on athletic ability? Intelligence? Eye color? Sex? Is there a line to be drawn here and, if so, where do you draw it?

For the record, and with no pretense whatsoever that this represents anything more than a personal, nonscientific decision, here's how I would decide the cases above: I would treat case 1 as an issue of public health. Just as we must warn and examine those at risk for tuberculosis because they've had contact with an infected person, we must tell the woman's siblings and children of their risk. In case 2, I would supply the genetic information and leave the decision to the couple, on the grounds that such decisions are too important and too personal for governments to interfere in them.

What Will Follow the Pill?

THERE AREN'T MANY technological changes that have had a more profound social impact than the introduction of oral contraceptives in the 1960s. Based on an old-fashioned technology, "The Pill" was quickly adopted in this country and around the world. It is surprising to realize, then, that since that initial burst of activity, there has been very little progress in contraception technology during the last twenty years.

It's not that there are no new ideas. In fact, as with every other aspect of biomedicine, our newfound ability to deal with living systems at the molecular level opens the way for new techniques. As is the case with antibiotics, however, the market forces operating in this area seem to have discouraged further research by pharmaceutical companies. There are successful contraceptives on the market now, and almost no incentives for companies to undergo the enormous expense (and accept the enormous legal risk) needed to bring new ones to market. In fact, the only recent development is the drug RU486, developed in France, which induces menstruation and is in effect an abortion pill.

This is not to say, however, that the field is completely dead. In fact, a certain amount of effort has gone into developing male contraceptives and "morning after" drugs that can be taken to prevent conception but that do not necessarily interfere with the menstrual cycle. The most successful male contraceptive so far developed involves injections of testosterone. While this seems to be effective in reducing sperm counts, it requires injections every few weeks, which means that it will surely not find a widespread market.

Most of the research on contraception today targets a critical step in the process of fertilization — the fusion of sperm and egg. At the molecular level, a number of steps must be completed for fertilization to occur. First, the molecules in the outer membrane of the sperm have to "recognize" and fit onto particular molecules of the outer layer of the egg — the sperm has to "dock" on the egg. Once the docking has occurred, the sperm releases enzymes that create a small hole in the outer covering of the egg through which the sperm can move. Finally, certain enzymes act to fuse the outer coatings of the egg and sperm. At the molecular level, then, there are at least three possible points of intervention: you can block the recognition process, the action of the enzymes that open the coating of the egg, or the final fusion. In fact, current research on contraception is exploring all of these options.

In 1980 scientists identified the molecules in the outer coating of the egg to which the sperm binds. As is the case with all molecular activity in living systems, this binding depends on simple geometry — a lock-and-key fit between the molecules. The contraceptive strategy is to make synthetic molecules of the same type and shape as those on the outer layer of the egg. These free-floating molecules will then lock on to the molecules in the coat of the sperm. In effect, by flooding the system with synthetic keys, we can keep the sperm from docking with the real key in the outer coating of the egg.

The strategies for dealing with the sperm's release of enzymes that allow it to penetrate the outer layer of the egg are slightly different. We don't yet have a clear picture of what's going on at the molecular level, but it seems that part of the process involves calcium chemistry, particularly channels in the egg's outer coating that allow calcium to move into the sperm. There has been some evidence that blocking these calcium channels — in effect, keeping calcium from entering the sperm — seems to prevent the union of egg and sperm, at least in the laboratory. At present, scientists are trying to identify the genes that code for production of the enzymes that control the calcium channels, with the ultimate goal of producing a "designer drug" that will keep those genes from being activated.

Finally, scientists are starting to look at the last stage of fusion — the fusing of the outer membranes of the sperm and the egg. Specific proteins apparently trigger this reaction, and, as with the enzymes that open the way for the sperm to get into the egg, scientists are thinking about a designer drug that will block their action. Of the three approaches, this is the most speculative.

It does seem strange, given the enormous fund of techniques that have been developed in biomedicine, that so little has changed in the science of contraception. Given the time it takes for new products to get to market, it is extremely unlikely that any product based on the research discussed above will be available commercially before the year 2010. Thus, research in this area will not contribute to limiting human population in the near term.

Is There a Future for Designer Fat?

 DESPITE THE HIGH levels of health consciousness that have developed in the United States in
the last decade, Americans are continuing to gain
weight. Not to put too fine a point on it, we are
getting fat. Many industries are based on the American desire
to weigh less, from the publishing of diet books to the manufacturing of exercise equipment. And researchers are trying to
understand the molecular basis of the problem and engineer a
molecular solution.

There is an evolutionary reason why modern human beings
find it so difficult to cut down on the amount of fat they take
in. For most of our history as a species, we lived in a hunting-
gathering society and very seldom had access to fats or sweets.
On the other hand, the ability to store fat against times when
food was scarce had an obvious evolutionary advantage.
Therefore we entered the modern age of affluence with bodies
that both craved fat and were very efficient at storing it.

An engineer looking at this problem will ask a two-part
question. Is it possible to substitute other materials for fats in
food, and if not, can we manipulate the fats so that the human
body does not store them easily? As it happens, both lines of
research are being pursued diligently by large food companies
in the United States.

Of the two lines of approach, the first seems to me the more
difficult. The problem is that fat affects both the texture and
the smell of food, and our sense of taste is very closely linked
to our sense of smell (remember how bland all your food
tastes when you have a cold?). Fat is related to the sense of

smell because most of the molecules that give foods their distinctive aromas are dissolved in fat and are released as the food is eaten. In addition, fat has a distinctive, slippery (dare I say slimy?) feel in the mouth that people crave.

Companies trying to produce substitutes for fat generally try to find molecules that will provide the texture and taste of fat but without as many calories as fat has. Two classes of substitutes are in commercial use — carbohydrates and proteins. Both, roughly speaking, have about four calories per gram while fat has nine. Carbohydrate substitutes are gums or starches from plants. When eaten, they tend to attract water and clump into little balls, giving the food the slippery texture of fat. The protein substitutes, generally from dairy products, also form little balls that give the food a smooth, creamy texture. Obviously, neither protein nor carbohydrate fat substitutes have zero calories — they just provide fewer calories than fat does.

It requires a fair amount of knowledge about molecular structure to make a fat substitute that the body won't absorb. The idea here is to modify the molecular structure so that the fat still has the right taste and texture and perhaps even the ability to withstand the high temperatures used in frying — and is undigestible by the human body. To understand how these substances work, you have to understand a little bit about the molecular structure of the fats most common in our diet.

A molecule of the stuff we usually refer to as fat would be called a triglyceride by an organic chemist. Each molecule looks like a capital Y, whose three arms are structures called fatty acids. It is these arms that carry the material we want to avoid digesting. The center of the Y, where the three arms come together, is an alcohol called glycerol. When this complex Y-shaped molecule gets into the intestines, a particular enzyme in the digestive system attaches to the glycerol and breaks the arms off. The shortened arms are then digested by other enzymes.

Triglycerides break down because the first enzyme mentioned above is a good geometrical fit to the glycerol at the center of the molecule; if we can block this fit we can prevent digestion. The molecules now being engineered to substitute for triglycerides have a glycerol at the center but have anywhere from six to eight arms instead of three. The added arms change the shape of the center of the molecule enough that the ordinary human digestive enzymes cannot get at it. The molecule simply passes through the digestive tract without being taken up by the body. It delivers, in other words, zero calories.

I suspect that this sort of molecular engineering is the wave of the future in food technology. Rather than expecting people to watch their diets, food companies will provide foods that satisfy the cravings we all have but with molecules modified so that they cannot be absorbed by the body. In 1996, in fact, the Food and Drug Administration approved the first such fat substitute, a product called Olestra, for use in snacks. You may recall that there was some controversy about the decision because this sort of substitute removes some fat-soluble vitamins from the body and, in a few cases, seems to cause diarrhea in users.

And who knows — perhaps one day when the waiter tells you that the cheesecake has no calories, he may actually be telling the truth.

Can We Live Forever?

 DURING THE Bronze Age, you could expect, at birth, to live eighteen years. By the Middle Ages this expectation had climbed to thirty-three years. At the beginning of the twentieth century, Americans had a life expectancy of forty-nine years, and today it is over seventy-five years. Advances in technology, sanitation, and medicine have increased the average human life expectancy dramatically, particularly in this century. The obvious question — one that has more than passing interest for us all — is whether life expectancy will continue to increase in the future. With new developments in technology, is there a chance that individuals will live significantly longer than they do now — perhaps even achieve something like immortality?

There is one fact you need to know to understand the debate on this question. Though average life expectancy has indeed risen rapidly in historical times, we have no evidence that the maximum age humans attain — what we can call the human life span — has changed at all. The oldest reliable age for individual humans has remained at about 110 years throughout recorded history.

In fact, over the past few decades serious research has tended to explode the myths of individuals living to 160 years and more. Such claims have been reported from a few extremely isolated rural areas (in the Caucasus, for example) where there are no reliable birth records. In these regions, self-reported (but unreliable) ages in excess of 120 years are common. My favorite story to illustrate why these numbers are no longer accepted concerns a researcher who made two visits, five years apart, to an Ecuadorian village. On the second visit, he found that the age of local "centenarians" had increased by seven years!

The apparent conflict between increasing life expectancy and fixed life span can be understood if you think of human life as analogous to an obstacle race. Everyone starts the race running at the same speed, but some runners drop out along the way as they hit the obstacles, and only a few make it to the finish line. For human beings in their natural state, the obstacles include predators, starvation, disease, weather, and accidents. Technology and medical science have removed or lowered many obstacles in the race, so more and more runners make it to the finish line. In a perfect world, every runner who started the race would reach the finish line, and every human being born would remain vital and healthy until he or she reached the age of 110.

Two questions come to mind from this picture. First, is it inevitable that humans must become less vigorous as they age? Second, is the life span limit of 110 years fixed or can it, too, be pushed back by advances in science?

Of these two questions, only the first has begun to be seriously studied. Although the field devoted to the study of aging, called gerontology, is (ironically) in its infancy, we can make a few statements. First, it appears that all mammals go through an aging process if they live long enough (though in the wild they seldom do). From an evolutionary point of view, aging is an irrelevant process. Once an animal has reproduced, it is, evolutionarily speaking, of no further use, and natural selection can have no effect on it. Like a car that goes on running past its warranty date, the animal just keeps on going until it breaks down. In this view, longevity is an accidental byproduct of whatever characteristics an animal has that allows it to survive until it reproduces.

In this perspective, the goal of medical science is to keep the human body running as long as possible — to supply by art what has not been supplied by nature. The phenomenal increase in life expectancy is one result of this attempt, and there is no reason to expect that the increase will be less phenomenal in the future. Right now the leading cause of death in the United States is cardiovascular disease. If it could be cured or prevented, scientists estimate that life expectancy would increase by at least seventeen years, to ninety-two years. Cures for cancers and other killers could raise it even higher.

It seems, then, that we should start to think about the other limit to human life — the finite life span. Little work has been done in this field, but already you can see two camps forming — call them the optimists and the pessimists. The optimists, buoyed by advances in molecular biology and molecular medicine, argue that once the last few obstacles like cardiovascular disease have been swept away, science will begin pushing back the life span limit. After all, they say, all processes in the body are molecular, and when we can control molecules, we will be able to control everything, even aging. The pessimists, on the other hand, argue that aging will turn out to be just too complicated to control, and even if it isn't, the social consequences of unlimited life span would be just too horrible to contemplate.

I have to say that I find myself squarely on the side of the optimists, at least in principle. If life is really based on chemistry, then we ought to be able to change any living system by changing its chemistry. If that statement applies to curing diseases, I see no reason why it shouldn't apply to aging and life span as well.

How Does the Brain "See"?

TAKE A MOMENT to look around you. What do you see? Perhaps you see a room with colored walls, pictures, doors, and windows. Whatever you see, though, one thing is clear — a collection of cells in your eye and brain has converted incoming light from the environment into a coherent picture. Over the past few years, scientists have started to construct an amazingly detailed picture of how that process works.

It begins when light enters the eye and is focused on the retina at the back of the eyeball. There, in the cells called rods and cones (because of their shapes), the energy of the light is converted into a nerve signal. At this point, the external environment ceases to play a role in the process, and the mechanisms of the brain and the nervous system take over. The central question: how are those initial nerve impulses in the retina converted into an image?

Since the early twentieth century, scientists have known that the basic units of the nervous system are cells called neurons. Each of the many types of neurons has a central cell body, a collection of spikes (called dendrites) that receive signals from one set of neurons, and a long fiber called an axon through which signals go out to another set. The neuron is an all-or-nothing, one-way element — when (by a process we don't understand) it gets the right mix of signals from its dendrites, it "fires" and sends a signal out along its axon. The problem for brain scientists is to understand how a set of cells with these properties can produce images of the outside world.

The first bit of processing of the data carried by the light occurs in two layers of cells in the retina (oddly enough, these cells are located in *front* of the rods and cones, blocking incoming light). These cells are connected in such a way that a strong impulse will go to the brain from one set of cells if it sees a bright dot with a dark surround and a weak signal from that set otherwise. Another set of cells will send a strong signal if it sees a dark dot with a white surround and a weak signal otherwise. The signal that goes to the brain, then, consists of impulses running along axons that, in effect, break the visual field down into a series of bright and dark dots.

These signals arrive at the primary visual cortex at the back of the brain, and from this point on the system is designed to build the collection of dots back into a coherent picture. The signals feed into a set of cells in one particular layer of the visual cortex, located in the back of the brain. Each of the cells in that set will fire if it gets strong signals from a certain subset of cells in the retina. For example, one of the cortex cells may fire if it gets signals corresponding to a set of dark dots tilted at an angle of 45 degrees. In effect, this cell "sees" a tilted dark edge in the visual field. Other cells will fire for bright edges, for edges at different angles, and so on. The output from these cells, in turn, is passed on to other cells (perhaps in other parts of the brain) for further integration.

So the process of seeing is far from simple. Neurons fire and pass information upward through a chain of cells, while at the same time feedback signals pass back down the chain and affect lower cells. Working out the details of this complex chain is one important area of research. Another is following the chain upward, to ever more specialized neurons. There are, for example, neurons in the temporal lobes (located on the sides of the brain) that fire only when they get signals corresponding to certain well-defined patterns — one set of cells may fire strongly in response to a dark circle with a horizontal line through it, another to a star shape, another to the outline of a square, and so on. Presumably neurons higher up the chain combine the output of these cells to construct even more integrated versions of the visual field.

Thus our image of the external world is built up in the brain by successive integrations of visual elements. Where does the process end? At one point, some neurophysiologists talked about the existence of the "grandmother cell" — the one cell in your brain that would fire when all the stimuli were right for an image of your grandmother. While this simple notion has fallen out of favor, it seems likely that within the next decade scientists will be able to trace the physical connections from the cones and rods all the way to the cells in the brain that put together the final picture. Other scientists are already trying to locate the cells in the frontal lobe that fire when short-term memory is activated. This work starts the next step in understanding the visual process — finding out how we recognize an object whose image has been constructed.

But even when all these neural pathways have been traced out and verified, the most intriguing question of all will remain: what is the "I" that is seeing these images? To answer *that* question, we'll have to know about more than neural circuits — we'll have to understand consciousness itself.

What Is the Connection Between Mind and Brain?

 THIS ISN'T PURELY a scientific issue; it's also a philosophical question, and an ancient one at that. But over the next few decades, as scientists unravel the workings of the brain in greater and greater detail (some of this research is discussed elsewhere in the book), the question of the connection between our brains and our minds will become more and more pressing.

The brain is a physical system. It contains about 100 billion interconnected neurons — about as many neurons as there are stars in the Milky Way galaxy. It's not the sheer number of cells that's important here, but the connections between them. Each neuron may receive signals from thousands of others, and may, in turn, send signals out to thousands more. The neurons seem to be arranged hierarchically: those that receive signals from the senses process them and pass them on to higher systems of neurons. In the end, by mechanisms we still haven't worked out (but we will do so!), these signals are converted, by neurons in different parts of the brain, into the final signals that produce images or smells or sounds. Thus the brain works basically by passing information from neuron to neuron. The most we can ever hope to achieve in any given situation is to follow the exact plan of that information. "Success" in brain science, then, would consist of a detailed knowledge of what neurons are firing in any circumstance.

The mind is . . . well, what is it, exactly? Formal definitions usually mention something like "the sum of mental activities," but that doesn't tell us much. On the other hand, we all have had the experience of mind. Close your eyes and think of an episode from your childhood. You probably can conjure up a fairly detailed visual image of some setting, maybe even some sounds and smells. You have these images "in mind," but where, exactly, are they? They obviously don't correspond to any sensory input into your brain right now, even though they must involve the firing of neurons somewhere. In the mid-twentieth century, in fact, neurosurgeons operating on the brains of conscious patients discovered that they could evoke these sorts of images by applying electrical current to specific parts of the brain (the brain has no pain sensors, so this is a completely painless procedure). Obviously, there is some sort of connection between the activity of neurons and our experience of "mind."

But what can that connection be? Who is the "I" that says "I remember," and where is he or she located? One way of looking at this question (which is almost certainly wrong) is to imagine that somewhere in the brain is an "I" who is watching the final products of the processing of signals by neurons. The essence of this view is that there is something in "mind" that transcends (or at least is distinct from) the workings of the physical brain. The seventeenth-century French philosopher and mathematician René Descartes ("I think, therefore I am") advocated such a view of mind/body dualism, so the hypothetical place where mental images are viewed is often referred to as the "Cartesian Theater." Think of it as a small stage located somewhere inside the brain where all of your experience plays itself out. (Descartes himself thought that it was in the pineal gland, since that gland is located in the center of the brain.)

If this view of mind as something that transcends the brain is not correct, then the firing of neurons is all there is. But this won't work either, and you don't have to be a neurosurgeon to see why. Suppose that at some point in the future a neuroscientist could say, "When you see the color blue, this particular set of neurons will fire in this particular order." Suppose that every time you saw blue those particular neurons fired and that they never fired in precisely that way when you saw anything else. Clearly, you would have established a correlation between the experience of seeing blue and a particular process in the brain.

But you would *not* have explained the experience itself! You are not, after all, aware of neurons firing — you're aware of the color blue, and the most brilliant neurological research in the world can't bridge that gap. Ancient philosophers, who had no idea what a neuron was, referred to this as the problem of *qualia* — a property, such as blueness, considered independently.

To understand the experience itself, you have to start in a wholly different branch of science — psychology. A lot of work is being done to understand human perception through such phenomena as optical illusions, which can illuminate odd corners of the mind, and through the process of learning itself. I get the impression that there are two groups of people trying to bridge the gap between mind and brain. On the one side, working up from the smallest scale, are the neuroscientists. On the other, working down from the largest scale, are the psychologists. Whether the two will ever come together is, to my mind, very much an open question.

How Many Species Are There on the Earth, and Why Does It Matter?

WE LIVE in an age when the disappearance of species is taken very seriously. It may come as a surprise, then, to find out that we really don't know how many other species share the planet with *Homo sapiens.* Wouldn't you expect that to be one of the facts we know about our world?

One problem is that until quite recently no one was keeping a running total of species as scientists traveled around the world cataloguing wildlife. Consequently, even the number of species that have been discovered and are presumably catalogued somewhere in the scientific literature is uncertain — it's generally reckoned to be between 1.5 and 1.8 million. Another problem is that insects represent by far the greatest number of species on the planet, particularly tropical insects, and they are some of the least studied animals.

Different scientists, starting from different assumptions and with different calculational methods, have arrived at wildly different estimates of the total number of species. Here are some samples of how these calculations are done:

Discovery Curves

If you look at the history of species discovery for birds, say, you see a pattern — only a few species are known at first, then a rapid growth in the number of species is followed by a leveling off as new discoveries become rare. By 1845 about 50 percent of the present 9,000 species of birds had been catalogued, and new discoveries became sparse by the early twentieth century.

If you assume that the shape of the curve that describes the discovery of new species is the same for other animals, then look at the discovery curves for each group up to now, you will estimate that 5–7 million will eventually be discovered.

Species Ratios

For groups like birds and mammals, which are pretty well studied, there are about two species in the tropics for each in the temperate zones. If this ratio is the same for insects, the total species number will be between 3 and 5 million.

Body Size

It seems to be a general rule that the smaller the animals are, the more species there are. There are, for example, more rat-sized animals than elephant-sized ones. The general rule seems to be that each time the size drops by a factor of 10 (from creatures a foot long to creatures an inch long, for example) the number of species increases by a factor of 100. If this rule is found to hold down to animals a hundredth of an inch long, the total number of species on earth is about 10 million.

Inferences from Small Data Bases

The idea here is to count every species in a small area (for example, a single tropical tree), then assume that the ratios of insects to plants and other species are the same worldwide. (This is a very shaky assumption, but with our current level of ignorance, it's a start at least.) Here's how one such calculation works: 160 species of beetle are counted in the canopy of a tropical tree. Worldwide, there are about 60 species of non-beetle insects for every 40 species of beetle. If this ratio holds for that particular tree, then there must be 240 species of non-beetle insects in the canopy, for a total of 400 insect species. Scientists estimate that about two-thirds of the insects live in the canopy, one-third in other parts of the tree. This gives us 600 species of insects in the entire tree. There are about 50,000 species of trees in tropical forests, and if each contains 600 insect species, then there are at least 30 million species of insects in the world.

Each of these methods (particularly the last) involves a long string of assumptions that may or may not be correct. The only way to improve the estimates is to do a lot more field work. Over the next decade, I expect we will see a number of detailed studies of small areas of tropical rain forests, studies that improve the estimates of ratios of insects to other animals to plants that are used in these calculations. I also expect we will see more studies that clarify the ratios of tropical to nontropical species. I do *not* expect that we will ever have a complete census of all the species on our planet. Given the rapid changes taking place in every ecosystem, I'm not sure how worthwhile such a census would be in any case. I am sure, however, that over the next decade we will have much better estimates of the number of species on our planet than we do now.

These surveys will become more important as governments get more serious about cataloguing the environmental diversity in their territories, and they may well serve as guides for wildlife management policies. The total species number will then figure prominently in debates about worldwide efforts at conservation.

For what it's worth, my own guess is that the final tally of the world's species will come in at 5–6 million.

Are We the Men We Used to Be?

 THE SUBJECT of this item is a sleeper — one of a number of little known environmental problems that could, in principle, seize national attention at any time in the future. On the other hand, it could be a false alarm, and after you finish these three pages you may never hear of it again.

The basic situation is this: in the early 1990s, several European studies suggested that sperm counts have been dropping steadily since 1940, from about 113 million per cubic millimeter in 1940 to about 66 million in 1990.

Confronted with these data, we can ask two important questions: (1) are the numbers correct, and do they really mean what they purport to mean? and (2) if they do, what could have caused the trend that is seen in the data?

At first blush, one would guess there would be no question about what the data mean. After all, there are data on sperm counts in medical records from most of the developed world. The problem is that to make sense of the data, you have to put together studies done under different conditions by different people, using different criteria and microscopes for determining sperm counts. Over a fifty-year period, all of these factors can vary quite a bit. If you gave identical samples to a technician in Denmark doing a sperm count today and a technician working in Dallas in 1940, would they come up with the same answer? Some scientists argue, for example, that the criteria for counting sperm and the sampling techniques used vary considerably from country to country and over time, and that all or most of the apparent drop in fertility could well reflect nothing more than differences in instructions given to technicians.

In addition, in 1996 researchers in the United States reported on a twenty-five-year study of sperm samples given by men before undergoing vasectomies. They found no drop in the count at all, but did note large geographical differences — men in New York, for example, seemed to have higher counts than men in California. The researchers suggest that the European studies, which relied heavily on New York data to determine their early numbers, may have recorded differences in location rather than in time.

So that brings us to the second question: what could be causing a decline? The culprit seems to be a set of molecules that are widely used in industry, and which, in small amounts, can be found in the environment in most of the industrialized world. These organochlorides, which incorporate chlorine molecules into their structure, are important components of PVC plastics (widely used in plumbing systems) and of many common pesticides and herbicides. In fact, about two thirds of the plastics in the United States use chlorine compounds, as do about a fifth of all modern pharmaceuticals.

Many of these molecules have a shape that apparently allows them to mimic the female hormone estrogen when they are taken into the human body. As in all molecular phenomena, the actions of estrogen (which plays an important role in the biochemistry of both men and women) depend on its shape. It is taken into a cell because it is the "lock" that fits the "key" of the cell's receptor molecules. The explanation that has been proposed for the claimed drop in sperm counts is that certain organochlorides have just the right shape to fit onto those same receptors. According to this theory, the presence of these molecules in the environment adds to the levels of estrogen-like molecules normally found in (and produced by) the body.

Early in the development of the human fetus, genes may produce chemical messengers that, in effect, throw chemical switches that lead to the production of male hormones. If the switches aren't thrown, the fetus will be female. In the language of engineers, female development is the default mode for the fetus. Given the extreme importance of chemical signals during this crucial developmental stage, it is not unreasonable to suggest that extra estrogen-like molecules, perhaps delivered through the mother's bloodstream, could affect male development. There is some evidence from laboratory animals that some compounds act this way. On the other hand, critics of this theory point out that organochlorides occur naturally (seaweed, for example, is full of them), and ask why artificially produced molecules should have this effect when the same molecules produced in nature don't.

If we ever find a "smoking pistol" establishing the link between fetal development and organochlorides, we will undoubtedly have to remove them from the environment. Given the cost of doing so, however, we will have to be very certain of the link before we take action. This, I suspect, is where the debate on this issue will focus in the years to come.

What Is an Ecosystem Worth?

 AS HUMAN POPULATIONS grow and the environmental movement becomes more politically powerful, this question will be asked more and more. Should this wetland be preserved, or should it be paved for a shopping mall? Should this woods be cut down to make space for homes? These kinds of questions are, at bottom, questions about values. Is the ecosystem worth more to us as it is or as it would be if it were transformed?

At present, three different approaches are being taken to this problem:

What Would You Pay?

Social scientists often take a market approach to the question of the value of an ecosystem. A typical study might involve asking a sample of people, "What would you pay to keep area A as it is?" Answers are collected from a wide range of people and the data corrected for income, geography, and other variables (wealthy people are usually willing to pay more than are the poor, people near a proposed park may feel differently from those living far away, and so on). In the end, you get a number that represents the dollar value of that ecosystem as perceived by the general population. For example, if individual Americans generally felt that it was worth $100 to keep Yellowstone Park as it is, then the value of the park would be $28 billion ($100 times 280 million Americans).

This method of evaluation works well for recreational systems like parks, but it may not work so well if the ecosystem has functions beyond recreation. If people answering the question don't understand these functions, they won't be able to come up with a realistic value in their response.

Hidden Treasures

Virtually all modern medicines came, originally, from nature. Aspirin, digitalis, and codeine (to name just a few) are examples of plant-derived drugs in widespread use today. The "hidden treasures" argument uses this fact to argue that in the unexplored biota of areas such as tropical rain forests are drugs that would be of boundless benefit to humanity, and that the potential benefit derivable from these drugs has to be added to the value we place on an ecosystem.

From a purely biological point of view, this argument makes good sense. For hundreds of millions of years the process of natural selection has been at work in nature, refining chemical processes in all sorts of organisms. A single plant, for example, may produce dozens or even hundreds of different pesticides to protect itself from insects — and those insects will, over time, produce antidotes to the pesticides. Compared to the accumulated chemical expertise of a rain forest, Du Pont and Dow are small change. So at first glance, the argument for preserving ecosystems (most especially the rain forests) because of their hidden treasures makes sense.

I have to admit, though, that I find the argument unconvincing. In the first place it doesn't say that we should actually *preserve* any ecosystem. We would still have the advantage of the hidden treasures if we went in and preserved one copy of the DNA of everything in an acre of rain forest, then cut down the trees and paved it. More importantly, this argument ignores the fact (discussed elsewhere in this book) that pharmaceutical chemists will soon be designing our drugs rather than searching for them in nature. As this process gathers speed (and it's already well under way), the value of the hidden treasure will drop. I don't like arguments for social policy that are so clearly hostages to technological advance.

Unrecognized Services

When you exhale, the carbon dioxide from your lungs enters the air and, eventually, is taken in by a plant, which will replace the oxygen you used in that breath. The global ecosystem, in other words, purifies the air we breathe. Systems like the Florida Everglades do the same for the water that flows through them. The "unrecognized services" argument for preserving ecosystems rests on the fact that much of what we take for granted in the world (like air and water) is a product of the workings of the global ecosystem. Economists have never taken these factors into account in their calculations, the argument goes; if they did, they would see that ecosystem preservation is worth a lot more than their surveys show. How would you answer a question like "What would you pay to be able to take your next breath?"

I should point out, however, that accepting this type of argument would require a fundamental change in the way we approach environmental questions. Instead of talking about ecosystems and biodiversity as if they were important in their own right, we would be focusing on their value for one member of the ecosystem — *Homo sapiens*. In the end, however, I think the unrecognized-services argument is the only one that will garner enough public support for environmental preservation over the long haul.

The Attack of the Killer Bees

 OKAY, I REALIZE that in a book devoted to questions about the origin of the universe and the fate of humanity, an item about the introduction of the so-called killer bees into North and South America may seem a bit out of place. However, it is one example among many of what will be a major ecological issue in the future — the introduction of exotic species into new areas. The rabbits of Australia, the zebra mussels of the Great Lakes, and the kudzu of the American South are all examples of this phenomenon. A new species is introduced, either intentionally or by accident, and then, in the absence of predators, runs wild. And the story of the killer bees is as good an illustration of this process as any.

But the real reason I'm bringing up the subject is much simpler. Back in the 1970s, I bought an abandoned farm in the Blue Ridge Mountains. I built a house with my own hands, did the "back to the land" thing, and kept bees. Working with up to six hives, I got to know these creatures and entered into the ancient beekeeper's bargain: I provided them with shelter and a modicum of attention, and they provided me with honey. It's been ten years since the last of my bees took off for parts unknown, but I still have a few pounds of their honey hoarded in my cellar.

Humans and domesticated bees go back a long way; there are even paintings of beekeepers on the walls of Egyptian tombs. The ordinary brown and black honeybee, *Apis mellifera*, is probably descended from bees domesticated thousands of years ago in the Middle East. They came to the Americas along with European settlers and are now called European bees. They are social insects, living in hives centered around a single queen, who leaves the hive once in her life for a mating flight, when she seeks out a drone from another hive. Fertilized, she returns to the hive and lays eggs for several years before being replaced by a new queen. Beekeepers who want to control the breeding process usually go through a complicated process called "requeening," in which the old queen is removed from the hive and replaced with a queen that has been mated to a drone of known stock.

European bees have evolved to regulate their activities according to the length of the day. When days grow short, they start to shut down the hive, throw out the males, or drones ("We don't need *you* anymore, buddy"), and get ready for winter. When days start to lengthen, the queen starts to lay more eggs and the hive population starts to build up. In this way, there are plenty of workers available in the spring, when flowers are abundant. This pattern works well in temperate zones.

It doesn't work in the tropics, however. Flowers are most abundant during the rainy season, which doesn't necessarily coincide with the period of longer days, so the bees aren't as productive as beekeepers would like. In 1956 some beekeepers in Brazil decided to import forty-seven African queens to see whether they would produce more honey. In 1957, a visitor to the research apiary in Brazil removed from some hives the grid that kept the queens from escaping. No one knows who did this or why, but by the time the situation was recognized, some twenty-six colonies of Africanized bees had escaped (or, to use the beekeeper's colorful phrase, "absconded"). Killer bees, *Apis mellifera sceutellata,* entered the consciousness of the hemisphere.

As the bees interbred with local populations and worked their way northward — Venezuela in 1977, Panama in 1982, southern Texas in 1993 — several facts became apparent. Interbreeding produced hybrid strains with new traits, including a lowered propensity for producing honey and a greatly heightened propensity to be aggressive in defense of the hive. It was this last trait that gave them their nickname. All bees will defend their hives — I remember being chased several hundred yards by angry bees when I opened one of my more aggressive hives on a cloudy day. Africanized bees, however, are distinguished by the extent of their defense — they can keep attacking for hours, with stings numbering in the thousands. They are responsible for one documented death in the United States, and perhaps a thousand all told since their release.

As is the case with most introduced "exotics," the effect of the killer bees will be limited in the end. First, they will reach the limits of their natural range, and, second, humans will take steps to control them. Projections are that Africanized bees will be found in most of the extreme southern region of the United States by the year 2000, but they won't spread farther north because of the climate. In addition, beekeepers are starting to use two strategies to limit the spread of the invaders. One is requeening, which maintains genetic lines in domestic hives. The other is called drone flooding. In this technique, large numbers of European drones are released in areas where queens come to mate (these areas play roughly the same role for bees that indoor shopping malls do for teenagers). This increases the chances that new hives will not become Africanized. So in the end, the killer bees will be controlled. But how much easier it would have been if those hives in Brazil had never been opened!

6

Medicine
(Mostly Molecular)

Using Your Genes to Tell Why You're Sick

 WHEN I THINK about the state of medicine today, I am reminded of an insurance ad that used
to run in magazines some years ago. It showed
two people talking on a hillside while above them,
totally outside of their attention, a huge avalanche was descending. Like those cartoon characters, we, too, are about to
be engulfed in an avalanche, except that this avalanche consists
of knowledge and technology rather than rocks and snow.

In the 1950s the double helix structure of DNA was first un-
raveled. Since that time there has been an exponential growth
in knowledge about the basic chemical workings of the cell. As
I write this, discoveries are being made so quickly that before
long everybody will have to become a little bit of a molecular
biologist, just to be able to talk to his or her doctor.

Here's what we've learned about how living things work: if
you think of the double helix of the DNA molecule as a kind
of twisted ladder, then each rung of the ladder consists of two
linked molecules called bases. There are four possibilities for
each base — molecules called adenine, cytosine, guanine, and
thymine, abbreviated A, C, G, and T. The string of letters you
encounter as you go down the ladder is the genetic message
that runs your body and that you pass on to your offspring. In
humans, the message is about 3 billion bases long and is carried on structures called chromosomes in the nuclei of our
cells.

As I point out repeatedly in these pages (and I don't think
this can be said often enough), life is based on chemistry, and
the directions for running the miniature chemical factory we

call a cell are contained in DNA. Certain stretches of DNA, called genes, contain the codes for building the protein molecules that run the cell's chemical reactions, with each gene controlling one reaction. There are about 80,000 genes in human DNA, but a given cell uses at most a few thousand at any given time.

Human diseases often involve something going wrong with the chemical machinery in the cells — either some chemical reaction isn't taking place or it's not being controlled properly. This means that the condition can be traced back to problems in the DNA. For example, there may have been mistakes in copying bases here and there. Such "point defects" woud corespund to simple mispelings in a written sentece. Other mistakes involve copying the same sequence over over over over over until the cell's machinery is simply unable to cope. Yet others involve the omission of chunks of code. The point is this: each mistake in the code can give rise to a specific disease, and we now have the means to identify that defect in a sample of DNA. For example, no fewer than 450 different diseases, from gout to some rare congenital diseases I hope to God you never hear of, have been traced to specific locations on the X chromosome. This number will surely increase, since there are about 5,000 genes on this chromosome, each of which could, in principle, be involved in one or more diseases.

What's going to change your doctor's visits before too long is that a combination of robotics and simple chemical tests will make genetic testing as cheap (and probably as commonplace) as blood testing is today. The basic technique is simple: samples of suitably prepared DNA are dropped on a small grid. In each square of the grid is a chemical that will bind to a particular kind of miscopied DNA (for example, square 1 may bind to a particular misspelling of a given gene, square 2 to a particular kind of mistaken repeat, and so on). This test is set up so that the squares in which chemical reactions take place change in color or some other quality, while those that don't pick up a mistake are unchanged. At the end, a computer will simply read the grid and specify which sorts of anomalies are to be found in your DNA.

Today a standard blood screening in your doctor's office costs a bit over a hundred dollars and tests for about twenty different diseases. That's about five dollars per disease. People developing the kind of automated genetic tests I've just described think they can bring the cost down to about half that. There are, in other words, no technical or economic bars to this kind of technology, although, as discussed elsewhere, many ethical issues must be resolved before they come into widespread use.

As you might expect, the first use of such tests has been to confirm the diagnoses of people who already show symptoms of a disease. Eventually, however, routine genetic screening could identify individuals at risk for particular diseases. Genetic counseling is likely to be very useful to prospective parents, for example, alerting them to possible problems if they both carry the same defective gene.

In fact, this kind of screening is already being used in a most unusual setting. Among ultra-orthodox Jews, marriages are often arranged by matchmakers. Several matchmakers are already using genetic screening to see if a prospective bride and groom are carriers of Tay-Sachs disease, even though at the time of testing the two may not even have met each other!

What Will We Do When
Antibiotics Don't Work Anymore?

 A FEW WEEKS before writing this I had a rather unnerving experience. I swim routinely as part of my exercise program, and, as happens now and then, I woke up one morning with an ear infection. Nothing to worry about, I thought — a quick trip to the doctor and a course of antibiotics will fix it. When I started taking the pills, however, something frightening happened. *The antibiotics didn't work!* I had contracted a microbe that was resistant to what the doctor prescribed.

In my case the solution was simple — I was put on a different antibiotic and the infection cleared up. But what was only an inconvenience for me may well become a life-threatening hazard for everyone in the not too distant future. For the fact is that our current honeymoon with wonder drugs is coming to an end, and in the future we will have to work hard to keep ahead of the pathogens that constantly surround us.

The reason for the development of such antibiotic-resistant baceria is rooted in Darwin's ideas of evolution. In any population of microbes, there will be some in which the genes have undergone mutations, and some of these mutations may allow their carriers to survive an antibiotic attack. For example, penicillin works because it contains a molecule called beta lactam that blocks the formation of bacterial cell walls. Soon after penicillin was introduced, some bacteria were found that could produce a molecule that would break up beta lactam. For these mutants, penicillin was a godsend. Not only did it fail to kill them, but it killed off their competition, leaving them free to flourish. Over the years, scientists and microbes waged a deadly war. Scientists would develop a brand of penicillin with a slightly altered form of beta lactam, one that the resistant bacteria couldn't counter. This would work for a while, but eventually the bacteria would become resistant to the new drug, at which point the entire process would start over.

The widespread use of antibiotics in medicine and agriculture (livestock are in many cases routinely fed antibiotics to keep them healthy) has, over the past half century, led to the existence of a large pool of bacteria that are resistant to one or more antibiotics. The spread of drug resistance is enhanced enormously by the fact that bacteria of different species often swap genes with each other in a process known as conjugation. Thus if one strain develops resistance to a particular drug, others may develop it as well. For example, even though there are over a hundred different antibiotics on the market, at least one strain of *Staphylococcus* is now resistant to all but one.

Although it is scary to think of diseases like cholera and typhoid fever decimating human populations again, I have to say that I'm not all that worried about this threat. The main reasons we have fallen behind the bacteria at the moment are economic, not scientific. Drug companies simply haven't believed they could make a profit by introducing another new antibiotic. It is very expensive to bring a new drug to market, even after it has been developed in the laboratory, and up to now there hasn't been much need to expend that sort of money or effort. But as the need for new antibiotics has become clear over the past few years, the companies are starting to gear up again.

Several strategies are being tried. One is very straightforward — if the action of an antibiotic is blocked by a molecule produced by the bacteria, then add to the antibiotic something that will block the action of that molecule. For example, if we can find a molecule that will bind to the molecule that breaks up beta lactam, rendering it inert, penicillin will work just as it always has. This technique has the advantage of allowing us to bring back some of the "golden oldies" of the antibiotic wars. Some companies, for example, are talking about bringing back tetracycline by blocking the molecules that resistant bacteria use to pump the drug out before it can take effect.

This is an area where I expect "designer drugs" to play an important role. We already know a great deal about the structure of molecules that make bacteria drug-resistant, so at least part of the job of developing companion drugs for antibiotics has been done.

And finally, there is the hope of finding antibiotics that act in an entirely new way. Today's antibiotics work by blocking cell wall construction, by blocking the formation of proteins, or by blocking the action of DNA, but these are not the only possible routes of attack. In the 1980s a class of drugs was discovered in soil bacteria that worked in a totally different way. They were never developed or even analyzed to the point that we understood how they work, for the economic reasons cited above. But the mere fact that they exist tells us that we haven't yet run out of possibilities in the never-ending battle against disease.

Will We Find New Magic Bullets?

 IN THE JARGON of medical researchers, a "magic bullet" is a molecule (usually administered in the form of a drug) that blocks the action of disease-causing molecules without affecting anything else in the organism. One sort of magic bullet takes the form of a drug constructed to block the action of specific proteins. (Proteins are the molecules that, among other things, facilitate chemical reactions in cells — without them, the reactions don't happen.) That has been the area of the greatest effort in the search for magic bullets.

But lately there has been another approach to the search. Instead of trying to block the action of a protein after it has been made, some researchers ask, "Why not block the process that makes it in the first place?" This approach hasn't been carried as far as other magic-bullet searches, but a number of drugs based on it are already in clinical trial. Because the techniques depend on a detailed understanding of the human genetic system, they are sometimes referred to as "genetic medicine" (not to be confused with gene therapy, which operates in quite different ways).

To understand how magic bullets work, you need to know a little about the cell's machinery. The instructions for building proteins are carried in stretches of the DNA molecule called genes. There are about 80,000 genes in human DNA, and each gene codes for a single protein. The information on the genes is transcribed onto shorter molecules called RNA, with the RNA bearing the same relation to the gene that a photograph bears to its negative. RNA molecules are in a long string, with the information carried by a sequence of smaller molecules, called bases, that stick out from the string. There are four different bases, normally referred to as A, U, G, and C, for the first letters of their chemical names (adenine, uracil, guanine, and cytosine). Thus the message on RNA might read AAUG-GCU . . . In the body of the cell are a host of molecules whose function it is to read this message and use it to construct a protein.

One approach to blocking the production of a protein is to attack the RNA messenger that carries information from the DNA to the cell machinery. This technique, called "antisense," works as follows: scientists build a molecule that constitutes a small part of the "photograph" to which the RNA is the "negative." When these molecules are introduced into the cell, they bind onto the appropriate portions of the RNA. This renders the RNA unreadable to the cell's machinery, so the cell never gets the message that the protein is supposed to be made.

One important clinical trial of antisense techniques targets RNA that codes for a protein that is crucial in the reproduction of the human papillomavirus (the virus responsible for genital warts). Without this protein, the virus cannot reproduce and hence cannot spread within the body. If the clinical trials are successful, this drug will probably be the first of the new magic bullets to reach the commercial market. Other tests are under way for drugs to combat AIDS and a form of leukemia.

Another approach to genetic medicine, not quite so far along as antisense, involves looking at the first step in the protein production process, the copying of information from DNA onto RNA. Unlike antisense therapy, which blocks the action of RNA once it is made, these techniques block the production of RNA itself.

The stretch of DNA we refer to as a gene actually has several parts. One part contains the code for building the protein, but before that is a section that turns the transcription process on — a kind of switch, or control. When particular proteins fit themselves onto this region, the cell starts copying messages onto RNA.

The coding sections tend to be wrapped in proteins that protect them from other molecules that might be in the cell. The control regions, however, are not protected in this way. The idea, then, is to manufacture a molecule that will bind to the control region of a particular gene, preventing the special proteins from locking on and starting the transcription process. This is done by making a molecule that will wrap itself around the double helix of the DNA, blocking the proteins that would ordinarily fit there. In effect, a third strand wraps around the DNA along the helix, so this technique is often referred to as the "triplex" strategy. In the test tube, it seems to work against some cancer-causing genes. Whether the triplex molecules can be made in a form that will move through cell membranes, survive the hostile environment of the cell (which contains many mechanisms to destroy bits of foreign material), and attach to DNA in the nucleus remains to be seen.

Whether or not they succeed, both the antisense and the triplex approaches serve as excellent examples of how understanding the cell's genetic machinery allows us to fight human diseases.

How Does the Immune System Work?

 LIKE EVERY OTHER part of the human body, the working of the immune system, which protects us from many different kinds of threats, depends ultimately on the geometric shapes of molecules. As you might expect from something that has been evolving for millions of years, the immune system is complex and multifaceted. We understand the main outlines of how it works, but there are problems with explaining (and understanding) it in detail, for we are still learning about the molecular workings of the body. It is fairly easy to see how each piece of the immune machinery works in terms of molecules fitting together, but so many steps and processes are involved that any explanation tends to fade into a mist of strangely named molecules performing arcane functions — so be forewarned!

The central event in an immune response is recognition by the body of the "enemy," whether it is an invading virus or one of your own cells that has started to form a tumor. The first line of defense is supplied by a type of white blood cell called a B lymphocyte (the B stands for bone marrow, where the cell is made). Each B cell carries protein molecules, folded up into complex shapes, on its surface. Think of each of these molecules as a key. The B cell searches for a foreign object carrying a molecule that is the lock for that key — a protein on the outer membrane of a virus or bacterium, for example. When it finds the lock, the B cell starts to reproduce rapidly, producing a flood of molecules called antibodies, which bear the same key as the B cell. The antibodies bind to any other lock molecules they find, either blocking their function directly or attracting other components of the immune system to destroy them.

The B cells consist, then, of a large number of different "off the rack" molecules which then multiply when they come across their molecular counterparts. By some estimates, the human immune system is capable of producing more than 2 billion different keys, only a small subset of which actually come into play in a given individual. But once a particular antibody is produced, the B cells retain a memory of it — this is why one encounter with a disease like measles confers lifelong immunity and why vaccines work so well. Sometimes, however, this initial response weakens over time, which is why you need booster shots for many vaccinations and why a disease like chicken pox can recur later in life as shingles.

If an invading microbe manages to get inside a cell in your body, another kind of white blood cell, called the T cell (because it is made in the thymus), comes into play. In every cell, bits of protein are brought to the surface by a specialized molecule called a major histocompatibility complex (MHC). If the T cell recognizes the protein carried by the MHC as foreign, indicating that the cell has been invaded, it initiates a sequence of chemical events that result in the cell's destruction. The virus that causes AIDS (which is discussed elsewhere) attacks the T cells, thereby weakening the immune system and leaving the patient susceptible to many kinds of bacterial and viral invasions.

In the future, you can expect to see a lot more medical treatments that focus on the immune system. Over the past few years, for example, scientists have developed strains of genetically engineered mice that produce human antibodies. Perhaps in the future, large vats at "Antibodies, Inc." will cook up batches of human antibodies that can be injected directly to combat viral diseases. Researchers are also making enormously important strides in the development of vaccines.

Another promising area involves recruiting T cells to combat tumors. Some scientists believe that one of the main functions of patrolling T cells is to recognize and wipe out cancerous cells before they build up into tumors. In many cases, however, tumors provoke little or no immune response, apparently because the system includes a built-in safety factor. Before the T cell can go into action, it seems that it has to receive two signals from a cell: recognition of a foreign protein–MHC system, as discussed above, but also a signal from a separate molecule recognized by a separate receptor on the T cell. The second molecule seems to act as a stimulant to the immune system rather than as a direct actor in the immune response. In the laboratory, cancer cells into which these secondary molecules are inserted are subject to vigorous attack by T cells, and scientists are trying to use this finding to develop improved treatments for cancer.

I suspect that we will make advances in immune system medicine by first learning which molecules perform specific functions, then learning how to use that knowledge to cure diseases.

Why Doesn't Your Immune System Attack Your Own Cells?

 THE IMMUNE SYSTEM is designed to protect us from all manner of foreign invaders. To do its job, it produces a wide variety of cells, the most important being B and T cells (the letters stand for bone marrow and thymus, respectively, where the cells are made). These cells have structures called receptors on their outer surfaces that act as keys that fit locks on molecules of invading organisms. It is this molecular recognition that is the central event in the immune response.

Under normal circumstances, every cell in the body sports bits of protein on its surface, carried there by other specialized molecules. These displayed proteins play the same role in the body as a password does in the military. Passing T cells sense these proteins and leave the cell alone. We say the T cell "recognizes self." Autoimmune diseases like multiple sclerosis and rheumatoid arthritis occur when something goes wrong with the password recognition system and the immune system starts to attack healthy tissue.

One of the great problems in immunology has been to understand exactly how the immune system recognizes self — why it will attack cells in a transplanted liver, for example, but leave the patient's own liver alone.

The immune system generates billions of different receptors, and the immune response is triggered when one of them fits an invader molecule. This diversity of molecular keys is produced by a basically random mixing of bits of protein, so it is extremely unlikely that your immune system would produce *no* T cells that could attack your cells. However, the system tests the T cells before they are released from the thymus into the body. Because the thymus is replete with self cells, any T cell that sports a key for your cell's lock will inevitably bind to a nearby cell. If this happens, the T cell stops growing and eventually dies. Thus the only mature T cells that emerge from the thymus are those that do not react with self cells.

We do not know in detail why T and B cells that have been selected for their ability to ignore self cells suddenly start to attack them. One popular theory is that viruses trying to gain entry present to the immune system molecules that mimic those of the host's body. This process could prime the system to attack any cells bearing similar molecules, even self cells.

As is so often the case in modern medicine, dealing with autoimmune diseases requires a detailed understanding of what is happening at the molecular level. In multiple sclerosis, for example, B and T cells attack the sheathing that surrounds nerve cells in the brain and spinal cord. First, one set of molecules on the membranes of T cells sticks to corresponding molecules in the walls of blood vessels, triggering the production of proteins that create a small hole in the vessel wall. This allows the T cells to get through the vessel wall and into the brain. There they encounter triggering molecules on the surfaces of nerve cells and initiate the immune response. The net effect is that other cells from the immune system, as well as the T cells themselves, cooperate to destroy the sheathing around the major nerve cells.

Understanding the basic processes involved in MS at the molecular level has allowed scientists to begin to develop a multipronged attack on the disease. The basic idea is to take each step in the chain of events outlined above and find a molecule that will block it. One technique is to use drugs that block the processes that bring bits of protein to the surface of the nerve cells. In fact, the first approved therapy for MS (in

1993) involved the use of a drug called beta interferon, which prevents production of the molecules that carry the proteins. Without these carriers, there are no proteins for the T cell to recognize on the cell surface, and the disease is blocked.

Other strategies, now under development, would block the disease process at other points. For example, scientists are working to develop decoy molecules that would lock to the molecules that normally bring proteins to the cell surface, preventing those molecules from displaying self proteins that would trigger T cells. Another strategy involves finding molecules that will prevent T cells from sticking to the walls of blood vessels by making molecules that occupy the "sticky spots" on the T cells or on the vessel walls. Think of this strategy as covering up Velcro patches on the T cell or the blood vessel — patches that would otherwise stick to each other. Finally, we can try to develop antibodies that will attack some of the molecules that actually damage the nerve sheathing.

This long list of possibilities illustrates the way medicine is likely to proceed in the future. First we will understand the basic molecular workings of the disease process, then find molecular strategies for blocking it. I've concentrated on MS here, but I expect to see this strategy adopted across the board in the future.

What Causes Cancer?

 CANCER REMAINS one of the leading causes of death in the United States, and understanding how it works and how it progresses is clearly a high priority in medicine. We are now acquiring some of the necessary knowledge.

There are several questions about cancer that we have to explain. Why do some types run in families? Why do there seem to be stages of cell proliferation, from benign growth to the wildly growing metastasizing tumor? And why is cancer, by and large, a disease of later years?

To understand why a cell starts to divide without control, you have to understand first how cells know when to start and stop dividing under normal circumstances. Cell division is triggered by the arrival of proteins called growth factors, which fit into receptor proteins embedded in the cell wall. The receptor proteins, in response to the arrival of the growth factors, produce yet other molecules that trigger a set of reactions that ultimately cause the cell's DNA to produce proteins that participate in cell division. Thus a chain of signals both from outside (the growth factors) and inside (the molecules that eventually trigger the DNA) is needed to start the process of division.

At the end of the cycle, a similar set of events tells the cell to stop dividing. Molecules called growth-inhibiting factors are taken in by a different set of receptors in the cell wall, initiating the production of molecules that eventually cause the cell's DNA to produce proteins that stop the division cycle. These last proteins are often called tumor suppressors.

So cell division is regulated somewhat the way a car is. There is a "gas pedal" (the growth factors) to speed it up, and a "brake" (the tumor suppressors) to shut it down. Therefore, the process can go out of control in two ways. If there is a defective gene in the DNA, the cell may churn out the molecules that initiate cell division even if no growth factor is present outside the cell wall. This is like having the gas pedal on a car get stuck — the engine keeps racing even though no one is pushing on the pedal. Alternatively, the DNA may be damaged to the point that it cannot produce the molecules that shut the cycle down. This corresponds to a failure of the car's brakes.

In human DNA, every gene has two copies, called alleles — one from the mother, the other from the father. Each allele can produce the protein for which it carries the code. The control system can fail if one allele mutates so that it produces too much of a protein that triggers cell division — this is one way that the gas pedal gets stuck. What biologists call oncogenes (genes associated with the onset of cancer) are an example of this failure mode. Alternatively, both alleles can be mutated to the point where a necessary protein (such as a tumor suppressor) is not produced at all — a failure of the braking system.

If you inherit a defective gene from one of your parents, it doesn't mean that you are doomed to develop cancer because you lack tumor-suppressing proteins. As long as you have one functioning gene, your cells will be able to control the division cycle. It does mean, however, that if a mutation damages the good gene, there is no backup. This explains how cancer can run in families and why not everyone in the family develops the disease. What is inherited is an enhanced susceptibility because of the lack of backup genes.

Today scientists think about the development of cancer in terms of "clonal evolution" or, more colloquially, the "multiple hit" theory. The idea is that even with one or two mutations in crucial genes, the cell's control system is able to limp along. It takes several "hits" on various genes before cancer develops.

Scientists have recently worked out the sequence of hits needed to produce colon cancer. The first step is the loss of a "braking" gene on chromosome 5, which results in the development of a small benign polyp. The next is the mutation of a gene on chromosome 12 into an oncogene, which initiates cell growth to produce a large benign polyp. Finally, there are losses of suppressor genes on chromosome 17 and of a gene on chromosome 18 that may code for substances that help cells stick together. Only after all of these hits have accumulated will the growth become cancerous.

Over the next decade, a lot more of these detailed sequences will be worked out, and I suspect you will be reading a lot about the identification of specific genes involved in the development of particular cancers. This knowledge will have an enormous impact on cancer treatment. It will allow us to use genetic analysis to make better diagnoses, and, by identifying what protein is or is not present at a specific stage of the disease, it will allow us to design drugs to unstick the gas pedal or put on the brakes.

Have We Lost the War on Cancer?

 ON DECEMBER 23, 1971, President Nixon signed the National Cancer Act, "declaring war" on one of humanity's most feared diseases. A quarter century and more than $25 billion later, it's fair to ask what all this effort has bought us. As is often the case where science and public policy come together, the answer to this question isn't easy to find.

Understanding the status of cancer in the United States today requires information about death rates from the disease over long periods of time, which in turn requires data from the branch of science known as epidemiology. With some perseverance, one can collect information from death certificates and other records, but then you are confronted with two problems of interpretation: (1) do the data mean what they seem to mean, and (2) how are they to be interpreted?

If you just counted the number of deaths from cancer each year for the last fifty years, you would undoubtedly find a massive increase. But this number doesn't mean that we are in the middle of a cancer epidemic, because cancer is largely a disease of old age. A good fraction of increased cancer deaths result from the fact that the American population is aging. *This is good news!* More of us are living long enough to get cancer.

If you look at "age-adjusted" cancer rates, which correct for the aging of the population, you find that there has been about a 10 percent increase in cancer deaths since 1950. But even this number is deceiving, since almost all of it reflects a huge increase in lung cancer, which is caused by smoking. Take out the smoking effects and you find that since 1950 the death rate for all other forms of cancer combined has dropped about 15 percent.

That's good news, right? Well, not really, because the aggregate rates hide a very complex picture when we look at specific types of cancer. Cancers of the skin, prostate, and liver have increased dramatically, while cancers of the uterus, stomach, and bladder have dropped. In some cases, such as brain cancer, the picture is mixed — the rate has dropped slightly for people under sixty-five but gone up dramatically for those over sixty-five. Any discussion of cancer, therefore, quickly devolves into a detailed look at specific cancers and their causes.

One problem is that the dramatic improvements in treatment that have made headlines in recent years — death rates for childhood cancers fell by 50 percent between 1973 and 1989, for example — have occurred in types that weren't very common to begin with. Treatment of the leading killers — lung, colon/rectal, breast, and prostate cancer — has improved very little.

In the debate about how to proceed in the war on cancer, two different schools of thought are starting to develop. One school says that cancer is under control and that before long the new treatment techniques (discussed elsewhere in this book) will drive the death rates down. The other school sees the data as the first step in a major cancer epidemic brought on by pollution and environmental degradation. Unfortunately, the data just aren't clean enough to decide between these two interpretations.

The optimists argue that much of the apparent increase is due to better diagnostic techniques. For example, they say that before the 1980s, when MRI imaging became widely available, many deaths from brain cancer were attributed to stroke. In this view, the increase in brain cancer in the data is simply a result of better diagnosis, rather than an increased incidence of the disease. The optimists argue that better diagnostic techniques across the board have resulted in much earlier detection of cancers. In effect, next year's cancers are found today, causing a temporary skewing of the data that masquerades as an increased incidence of disease.

The environmentalists reject these sorts of interpretations. They point to differences in national cancer rates as evidence that environmental factors and diet can have powerful effects. In Japan, for example, breast cancer rates are about one quarter of what they are in the United States, but when Japanese women emigrate to North America and start eating a high-fat diet, they and their daughters start to contract breast cancer at the same rate as American women. They argue that more attention should be paid to the prevention of cancer through environmental cleanup and more healthful living.

My own sense is that preventing cancer by building a perfect world is not really a viable option, at least for the short term. We will have to wait for the new therapies to come on line before we can declare victory in the war on cancer.

Can We Make Antibodies to Order?

 TO COMBAT some diseases, we have developed vaccines to stimulate the immune system. For the great majority of serious diseases, however, either we haven't figured out how to do this or factors such as the rapid mutation of viruses make it very difficult to produce an effective vaccine. In these cases, the ideal procedure would be to make antibodies outside of the human body rather than wait for the immune system to manufacture them.

Two techniques are being developed to produce synthetic antibodies. In one, which involves the modification of mouse antibodies, the end product is called, appropriately enough, "humanized" antibodies. This technique is well along in clinical trials and may have reached the market by the time you read this. The second procedure, still in the research stage, involves creating libraries of antibodies from material taken from human immune systems. These are called "human" (as opposed to "humanized") antibodies.

To understand how these techniques work, you have to know something about the construction of antibody molecules. The general shape of an antibody is a capital Y, whose arms are made from two separate molecules, called the light and heavy chains; the two heavy chains bend and come together to form the lower leg of the Y. The ends of the two arms are the parts of the antibody that fit onto foreign molecules. Depending on the structures of the light and heavy chains, the ends of the arms can take many forms. Think of them as complex sets of loops, with each loop being a convoluted finger sticking out into space, ready to bind to a target site in an invading organism. The immense versatility of the immune system arises from the fact that each of these fingers can take many different shapes.

To produce a "humanized" antibody, various chunks of the Y in a mouse antibody are removed and replaced by the corresponding chunks from that antibody in humans. Those parts of the structure that vary greatly from one antibody to another in people are left in their mouse form, while those parts that seem to be roughly constant from one human antibody to another are replaced by their human counterparts. Such hybrid molecules can supplement the human immune system without triggering allergic responses, as mouse antibodies by themselves often do.

Humanized antibodies are in the advanced stage of clinical trials for use against a disease called respiratory syncytial virus (RSV). This is a flulike viral disease that hospitalizes about 100,000 children in the United States every year and is probably responsible for flu epidemics among old people in nursing homes as well. The virus has two different-shaped molecules in its outer membrane — one locks on to molecules in the membrane of cells in human lungs, the other fits onto other molecules and initiates the process by which the virus is taken into the cell. The antibodies being tested are designed to bind to the second of these molecules on the viral coat, so that even if the virus has attached itself to a cell, it cannot enter the cell and reproduce. RSV is a particularly good trial for this technique, because like influenza in general, it is seasonal. Consequently, one or two injections will confer immunity against the disease through the winter; we can hope to see children and seniors being given antibody shots much as they are now given flu shots.

The second kind of manufactured antibodies — human antibodies — are made by a rather different process. Material gathered from the sites in the body where the heavy and light chains of antibodies are made (the spleen, for example) is used to produce strands of DNA that code for the chains. Many such genes are gathered and then scrambled — that is, all possible combinations of different kinds of heavy and light chains are constructed. There can be millions of such combinations, each producing a different potential antibody, even in material gathered from a single individual. Each of these antibodies is then tested to see whether it will bind to the target against which we wish to make an antibody. For example, if we want an antibody against RSV, we take a large collection of these scrambled antibodies and simply dump them onto a plate containing molecules from the coat of the virus. Those that bind will stick to the viral molecules on the plate; the rest are thrown away. The result: we can produce genes for a large collection of antibodies effective against that particular virus. The DNA for those antibodies can then be spliced into standard bacteria and grown like any other genetically engineered molecule.

The development of synthetic antibodies will change what happens when you get sick with a virus. You'll go to the doctor's office, get a shot, and feel better. You'll never have to know about the technical virtuosity that made it all possible.

Will We Be Able to Cure AIDS?

 THERE HAS BEEN a long-standing debate in public health circles about the amount of money spent on AIDS research. Critics point out that annual deaths from AIDS worldwide run at about 550,000, which is small compared to other, more familiar diseases such as pneumonia (over 4 million), tuberculosis (over 3 million), and malaria (over 1 million). Figures like these caused many people (including me) to question the enormous AIDS research budget when other, more potent, killers are given short shrift. As I have watched the AIDS research effort unfold, however, I have changed my mind. The reason for this switch is that AIDS was the first major new viral disease to come along after scientists had achieved an understanding of the molecular workings of living systems. Since viral diseases are likely to pose a major threat (some would say *the* major threat) to future generations, the tools being developed in the fight against AIDS will not only alleviate current suffering but will also stand us in good stead when new (and perhaps more easily spread) killers arrive on the scene.

AIDS is caused by a microbe known as the human immunodeficiency virus, or HIV. One result of the research program so far is that HIV has become the most intensely studied virus in history. We have an astonishingly detailed picture of the life cycle of this virus, along with a list of "targets" that might give us a chance of finding a cure (or perhaps a vaccine) for the disease.

The virus is spherical, and from its outer coating project spikes composed of a set of complex molecules. These molecules happen to fit into corresponding molecules on the surface of a number of human cells, most importantly one kind of T cell in the human immune system. Like a soldier who approaches enemy lines equipped with a password, the virus is "recognized" by the T cell. The outer coating of the virus then melds with the cell membrane, allowing the inner core of the virus to enter the cell. This core is a truncated cone made of proteins that contain two short stretches of RNA (a molecule related to DNA) and three different kinds of proteins to facilitate the reactions that come next.

Once inside the cell, the inner core of the virus begins to dissolve. One set of viral proteins facilitates a reaction called reverse transcription, in which materials from the cell are put together to make stretches of DNA from a template on the viral RNA. These short stretches of DNA (containing a total of nine genes) migrate to the nucleus of the cell, where a second set of viral proteins integrates them into the cell's own DNA. At this point the cell's chemical machinery has been coopted by the virus and will start churning out stretches of RNA and other proteins, which, with the help of the third viral protein, come together to form new viruses that bud off from the cell membrane. Eventually, the resources of the T cell are exhausted by the production of new viruses and the cell dies. Deprived of T cells, the immune system falters.

The knowledge summarized in the last two paragraphs represents the fruit of an enormous research effort by the biomedical community. This understanding of the HIV life cycle suggests ways in which we might deal with the disease. There are basically two strategies: we can find a way to deal with the virus once it is operating in human cells (that is, find a treatment for AIDS), or we can try to induce the immune system to fight off the virus (that is, find an AIDS vaccine). Both strategies are being pursued actively.

Treatment strategies focus on finding ways to block critical steps in the HIV life cycle. For example, drugs like azidothymidine (AZT) work by blocking the reverse transcription that produces DNA. These drugs are similar enough to the molecules normally incorporated into DNA that they are taken into the chain, but once there they block formation of any further links. Alternative strategies, which have shown success in clinical trials, involve trying to block the insertion of the virus's extra genes into the cell's DNA or to prevent the HIV genes from functioning once they are inserted.

The major problem with all treatments (and vaccines, for that matter) is HIV's rapid rate of mutation. It appears that the reverse transcription process is subject to high rates of errors — each time the process is carried out there are three or four "mistakes" in the resulting DNA. This means that every virus that comes out of a cell will be slightly different from the virus that went in. That is why, after a patient takes AZT for a few months, viral strains that are resistant to it begin to appear in his or her blood, and it is why so many different strains of the virus are known.

The high rate of mutation makes it extremely difficult to produce vaccines against HIV. The object of research is to use some nonlethal part of the virus (for example, one of the proteins in the outer coating) to induce an immune response, so that the immune system will wipe out the virus before it gains a foothold. Like the common cold, however, HIV may change too fast for an effective vaccine to be developed.

Where Are the New Vaccines?

 THE BEST WAY to deal with a disease is never to get it. The human immune system, whose workings are discussed elsewhere in this book, is designed to protect the body from invaders, to wipe them out before they can gain a foothold. Since 1796, when Edward Jenner found that inoculating people with the cowpox virus would protect them from getting the much more deadly smallpox, many major diseases have been controlled by vaccination. And vaccination, in the end, depends for its effectiveness on the immune system.

Smallpox used to be a deadly scourge. It was smallpox, rather than military action, that killed the greatest number of Indians in North and South America after their first contact with Europeans. (In an episode that will never, I fear, be included in modern politically correct histories, Lewis and Clark took smallpox vaccine along on their epic expedition, with instructions to inoculate as many Indians as they could). Today the smallpox virus is the first pathogen to have been completely eliminated from the earth (the last samples, kept in scientific laboratories, are supposed to be destroyed soon). This is the power that a vaccine program, properly administered, can have.

All parts of the immune system — B cells, T cells, antibodies, and so on — do their work by geometry. That is, they identify their target because their molecules match the shapes of molecules on the invader. And after the immediate battle has been fought, a small number of the B and T cells carrying the "key" that was found to fit the "lock" on the invader continue to circulate in the bloodstream. These so-called memory cells provide a quick defense against further invasions. It is memory cells that confer lifelong immunity once you've had a disease like mumps or measles, and production of memory cells is the goal of vaccination.

Today the struggle to push back the limits of vaccine technology is being waged on three fronts: (1) basic research aimed at understanding the way the immune system works, (2) research to improve existing vaccines and develop them for more diseases, and (3) political strategies to deal with the demographics of infectious disease.

Although we understand some of the main features of the immune system, there are huge gaps in our knowledge. The system is extraordinarily complex, and we often cannot answer detailed questions like "Why (or how) does molecule X perform function Y?" For example, the purpose of vaccination is to produce memory cells, yet scientists argue vehemently about how memory cells are produced by vaccines and how long they last. This issue is more than just idle curiosity — the answers are crucial to developing effective vaccines. Certain vaccines are known to trigger the immune response, but they fail to confer long-lasting protection and, in a few cases, even make the subject more susceptible to the disease than he or she was before. We must learn why some vaccines fail in order to find better methods in the future.

It sometimes happens that the immune system lock fits more than one invading key. This is what happened with the original smallpox vaccination — the proteins on the cowpox virus were similar enough to those on smallpox that memory cells developed for cowpox were called into action when smallpox showed up. Many vaccines consist of viruses that have been killed but that still sport crucial proteins on their surfaces. Memory cells key onto these proteins in the dead virus, then attack live viruses that carry them.

One particularly exciting area of research involves identifying the exact protein in a virus or bacterium that the immune system should attack, then administering that protein by itself as a vaccine. Alternatively, genes for the protein can be inserted into a harmless virus, which then triggers the immune response. As with killed viruses, the memory cells will then attack the target protein when it is carried by an invader. The technique of keying the immune response to specific proteins, if it comes into commercial use, would eliminate many of the side effects and dangers associated with current vaccines. A killed virus might not be as dead as was thought, and therefore dangerous, but a protein by itself cannot cause a disease. In addition, the technique skirts another problem associated with the immune response. The immune system normally recognizes and responds to many of the invader's proteins, but only a few of those proteins may actually be crucial in developing the disease. By zeroing in on proteins that are essential to the invader's life cycle, the vaccine can be made much more effective.

Finally, I should note that the main problems for vaccination programs in the future are not scientific but political. The problem is that vaccines have been produced mainly in the industrialized countries, but the greatest need for them is in the poorer nations. Up until recently, drug companies that developed vaccines charged high prices in industrialized countries, in effect subsidizing lower prices in poorer nations. As health care costs go up, however, advanced countries are becoming less willing to bear these costs, and research on vaccines for major killers like malaria has suffered.

7

Evolution
(Mostly Human)

Where Did Modern Humans Come From?

AT THE END of the nineteenth century there was a major controversy in the sciences between seemingly unalterable conclusions from two important fields of science. On the one hand, theoretical physicists, with the kind of arrogance that only they can muster,* said that the laws of nature did not allow the earth and the sun to be more than 20 to 100 million years old. Geologists and biologists, on the other hand, argued that their data indicated a much older earth. In the end, this conflict was resolved when the theoretical physicists realized that (oops!) the newly discovered phenomenon of radioactivity changed their predictions so that now they, too, assigned the earth an age of billions of years.

Today, at the end of the twentieth century, a similar conflict seems to be brewing. This time the role of the brash theorists is taken by molecular biologists. Using the new, glamorous techniques of DNA analysis, they are insisting that all modern humans had common ancestors a few hundred thousand years ago. Paleontologists, meanwhile, point to their old-fashioned (and *very* unstylish) collection of fossils and say that the world just didn't evolve that way. The conflict can be characterized as a duel between the out-of-Africa school and the multi-regionalists.

In the out-of-Africa theory, *Homo erectus,* the precursors of modern humans, dispersed from Africa 2 million years ago but were then replaced by modern *Homo sapiens,* who left

*I am a theoretical physicist, so I'm entitled to make this observation.

Africa about 100,000 years ago. In this theory, human diversity is of fairly recent origin. In support of their ideas, these theorists point to family trees derived from analysis of DNA in modern humans from different parts of the globe. (These arguments are discussed elsewhere.)

The paleontologists, for their part, point to the sparse collection of human fossils they have scratched out of the ground over the past century. Those who rely primarily on these data also believe, like the out-of-Africa supporters, that *Homo erectus* came out of Africa about 2 million years ago and spread to the rest of the world. Unlike their colleagues, however, they hold that modern humans then evolved independently in those different regions. In its most extreme form, the multiregional hypothesis is almost certainly wrong — it is highly unlikely that identical evolutionary changes could have taken place in different environments. In a modified form, however, in which human beings developed separately in different regions of the world with some gene flow between them, the theory corresponds pretty well with known facts about fossils.

If you look at skulls in a particular region (Australia, for example, or Asia or Europe), there seems to be a smooth, continuous record of change in each area, with no evidence whatsoever of any sudden influx of new peoples or the replacement of indigenous populations by newly evolved humans from Africa 100,000 years ago. There just doesn't seem to be any way to reconcile these data with the evidence from DNA.

So now what? We have two bodies of evidence, each with seemingly impeccable credentials, that lead us to mutually contradictory conclusions — the same situation our predecessors faced at the end of the last century. If we believe the DNA, modern humans spread around the globe from Africa starting about 100,00 years ago. But if we accept that, we have to ignore the evidence of the fossils, and if we believe the fossils, we have to ignore the evidence from DNA.

As you may have guessed from the tone of my introductory remarks, I suspect that when this issue is finally resolved, it will be the molecular biologists who will wind up with egg on their faces. The time I have spent collecting fossils with paleontologists in the Big Horn Basin of Wyoming has given me an abiding respect for both the difficulty involved in obtaining fossil evidence and the plain, rock-hard reality of the data. As a theoretical physicist, on the other hand, I am skeptical of arguments, like those of the molecular biologists, based on long strings of theoretical assumptions.

In the end, I suspect that the conflict between the two schools of thought will be resolved when we understand the difference between the movement of genes and the movement of peoples. It is the first that the molecular biologists measure, the latter (I think) that is preserved in the fossil record. For example, one boatload of voyagers, blown off course, could easily introduce a whole new set of immune-system genes into an indigenous population without necessarily producing any noticeable changes in the shape of fossil skulls. In the end we will find that human history is much more complicated than either of the two theories, in their simplest forms, can accommodate. I suspect that the true history of the human race will be found to consist of many migrations and counter-migrations over the last few million years and will show a complex interweaving of human genetic material throughout recent history.

The Molecular Origins
of Human Beings

AS A SPECIES, human beings are remarkably similar to each other. There is less variation between the DNA of two human beings from opposite ends of the earth than there is between the DNA of two gorillas from the same African rain forest. Nevertheless, groups of humans do differ from each other, and from those differences scientists hope to understand how we came to be what we are.

There are several different approaches to dealing with this problem. One is to look at the fossil record of human beings. The second is to study the most underappreciated living fossil in the world — the human organism itself. Human DNA represents millions of years of accumulated evolution, and we are only now starting to learn how to decode it.

The basic technique is easy to describe, if not always easy to carry out. Sections of a specific stretch of DNA from different individuals are sequenced and analyzed for differences. The assumption is that the more differences there are, the longer it has been since those two shared a common ancestor. If you feed this information about many individuals into a computer, it will produce a family tree stretching backward in time to the point where all the individuals shared a common ancestor (the technical term for this is the "coalescence time").

The first attempt at this sort of analysis used DNA from structures in the human cell called mitochondria, which is carried in the egg and hence inherited only from the mother. The analysis led to the rather unfortunate hypothesis that all humans were descended from one woman (named, appropriately enough, Eve) — who was featured on the cover of *Newsweek* magazine. This hypothesis was based on faulty data analysis and has been largely discarded. Data on mitochondrial DNA is now being supplemented with studies of DNA from the nucleus of the cell. Let me tell you about one study of nuclear DNA so that you get a sense of how these studies work.

In human beings, DNA in the nucleus is segregated into twenty-three separate pairs of chromosomes, and on chromosome 6 there is a stretch of DNA, containing about fifty genes, that codes for various proteins in the human immune system. The gene for each protein is located in a specific spot on the DNA but may come in many different forms (much as the gene for eye color, for example, comes in many different forms). Some immune-system genes may appear in as many as sixty different forms — it is the ability to scramble proteins from these genes that gives the immune system its flexibility. As with the analysis of DNA from mitochondria, scientists compare nuclear DNA from two different individuals and look at the number of substitutions you would have to make to get from one to another. The more substitutions, the longer the coalescence time.

Genes in the human immune system exhibit a rather startling characteristic. It is possible to find more differences between the DNA in these specific genes from two human beings than between the DNA of either of those two human beings and a chimpanzee. When we feed data on immune-system genes into the computer, we produce family trees in which the last ancestor that had the same gene lived 30 million years ago, long before humans split off from chimpanzees. These genes, in other words, entered the genome before there was ever a human race.

It is possible to use genetic theory to estimate how many individuals have to be in a population to keep a gene with so many different forms intact — if the number of individuals was too small, it would be impossible for sixty different versions of a gene to remain in the population for more than a few hundred generations. From this, scientists conclude that for the last few million years, the human population has consisted of about 100,000 breeding individuals, with possible "bottlenecks" when the population may have shrunk to several thousand or so. (It's sobering to think that we may once have been an endangered species!)

Most studies of mitochondrial DNA (and some nuclear DNA studies) indicate that the last common ancestor of modern humans lived between 100,000 and 400,000 years ago, with each stretch of DNA giving a slightly different coalescence time in this range. That different stretches of DNA should give different answers to the question of common ancestry isn't surprising. For one thing, you would expect that different genes entered the population at different times in the past; the genes that make native Americans different from their Asian ancestors, for example, clearly developed quite recently. Furthermore, there is no particular reason for changes in DNA sequence to accumulate at the same rate in different parts of human DNA. Advantageous mutations in genes that clearly have a high survival value (in the immune system, for example) would be expected to spread through a population much more rapidly than genes for something relatively neutral, like eye color.

Nevertheless, the verdict on human origins from molecular biology is unequivocal: all modern humans shared a common ancestor a few hundred thousand years ago, and all human diversity has developed since that time. Unfortunately, this conclusion flies in the face of the fossil evidence, a fact that is discussed elsewhere.

How Did We Get to Be So Smart?

IN THE END, it is intelligence that makes human beings different from other animals. The functions that we normally associate with intelligence are housed in the thin, wrinkled outer covering of the brain, the cerebral cortex or gray matter. If you took the average cerebral cortex, and flattened it out, it would cover about six pages of this book. For reference, the cerebral cortex of a monkey would cover about half a page and that of a chimpanzee (our closest living relative) about a page and a half. The question of how we acquired intelligence can be looked at in two ways: we can ask *when* our ancestors acquired it, and we can ask *why* they acquired it. The question of when is easier to answer, because we possess skulls of human ancestors going back at least 4.5 million years. It is a relatively straightforward process to take casts of these skulls and determine brain size.

The brain size measurements show a steady overall increase in our ancestors' cranial capacity. One particularly important evolutionary step was the development of *Homo erectus* about 1.5 million years ago. The range of brain sizes in *H. erectus* overlapped that of modern humans. This increased brainpower was reflected in new behaviors — the development of advanced tools, the domestication of fire, and the use of caves for shelter, to name a few.

The noted paleontologist Richard Leakey has this to say on the subject.

When I hold a *Homo erectus* cranium . . . I get the strong feeling of being in the presence of something distinctly human. . . .

[It] seems to have "arrived," to be at the threshold of something extremely important in our history.

Frankly, I tend to put more faith in this kind of intuitive leap by a scientist who's spent a lot of time working with fossils than I do in convoluted arguments based on skull measurements. For me, at least, the "when" has been answered — about 1.5 million years ago, something happened in our ancestors' brains that made them markedly different from those of other animals.

But what was that "something"? This is where we run into uncertainties, because the rules of the evolutionary game are that *every* stage in an organism's history must confer an evolutionary advantage, not just the final stage. It's no good saying, for example, that if an animal develops wings it will be able to escape its predators. You have to be able to show that having half a wing (or a quarter, or even just a stub) confers an evolutionary advantage as well. Thus, although it is obvious that having full-blown intelligence has been a great asset to our species, scientists still have to show how each step on the road to intelligence was dictated by the laws of natural selection.

In situations like these, evolutionary theorists look for overlapping functions — for example, some have argued that the stubs that later developed into wings were useful for heat dissipation, and that purely by chance the stubs turned out to be useful for rudimentary flying. In the same way, scientists now argue that in the history of developing human intelligence parts of the brain developed because they were useful for one reason and then, by chance, were found to be useful for something completely different.

Most theories of intelligence center around developments that followed from the freeing of our hands by our upright posture. A human with hand-eye coordination that allowed him to throw things more accurately than his fellows would clearly have a food-gathering advantage and be likely to have well-fed children who would grow to healthy adulthood. Thus you would expect the genes (and corresponding brain structures) responsible for hand-eye coordination to spread throughout the population. Some scientists argue that hand-eye coordination is tied in part to the brain's ability to deal with sequences, particularly novel and unfamiliar sequences. (To hit a piece of flint in the right way with a stone tool, for example, you have to plot out the sequence of hand movements before they actually begin — there's not enough time for corrections once the stroke starts.) These same planning abilities are also crucial for the development of language — indeed, there is some evidence that stroke victims who have lost some speech functions also have difficulty in learning to move their arms in new sequences. People who propose this sort of theory argue that the development of language centers in the left lateral area of the brain were an unintended consequence of improvements in hand-eye coordination.

As we learn more about the workings of the brain, I expect that we will see more explanations like this for various mental facilities, particularly those facilities we regard as uniquely human. The ability to deal with sequences, for example, is clearly essential to both long- and short-range planning, either of which could have developed as an unintended result of something else entirely. How many other attributes that we consider uniquely human arose as unintended consequences of these sorts of mundane skills? That, I think, is the central question.

Who Were the Earliest Humans?

MUCH OF THE evidence about our human ancestors comes from the discovery and study of fossils — replicas in stone of parts of skeletons. The real problem with tracing the human family tree is that we don't have very many fossils of our ancestors. There are lots of reasons for this. For one thing, early humans weren't all that common — they were probably no more populous than chimpanzees are today. In addition, bodies of land animals tend to get scattered by scavengers or decompose rather than be buried. And even if a fossil forms, there is no guarantee that after millions of years the rock in which it lies will be accessible. The most important fossils in the world may be out of our reach — buried 200 feet down under a suburban shopping center, for example. Another reason for the scarcity of human fossils is the method scientists use to find them. It often amazes people to learn that in our high-tech information age, fossil-hunters still walk along as they did a hundred years ago, eyes downcast, until they find a fossil lying at the surface.

This means that human paleontology is a field that has always been — and most likely always will be — starved for data. It's not like physics, where you can keep running an experiment until you accumulate all the information you need. For example, our knowledge of our most recent relative, Neanderthal Man, is based on about a dozen skulls and bits of fossil from perhaps a hundred individuals. This means that human paleontology is one of the few scientific fields in which a single new find can produce a complete revolution in our thought.

For example, in 1974 Donald Johanson, looking for fossils in Ethiopia, discovered the fossil we now call Lucy, at the time the earliest known human relative. Most family trees drawn since then show *Australopithecus afarensis* (the technical name for Lucy and her kin) at the base of the tree about 3.8 million years ago and various related species coming forward in time to the present. But since each member of the tree is known by only a few fossils, it is difficult to say exactly when each line appeared and died out, and whether a particular fossil represents a side shoot or the main trunk of the tree. Even today there is a debate about whether Neanderthal should be considered our ancestor or just a collateral cousin.

Nevertheless, when it comes to our more recent ancestors, we can trace out a rough outline of the family tree. It's when we try to link the most recent branches to the more distant past that we run into problems. We have reasonably good fossil records of primates up to about 8 million years ago. We know of a handful of primates from the earlier period — small, apelike creatures that few of us would have difficulty in separating from the human family tree. Between these early primates and Lucy, however, stretches a paleontological *terra incognita* 4 million years long. Not only are there virtually no primate fossils from this period, but rocks that might contain such fossils are quite scarce on the earth's surface. This is why the region in southern Ethiopia where Lucy was found is so important — it's one of the few places where you can hope to find early humans and their precursors. Unfortunately, this area has been politically unstable in the recent past and therefore has not always been open to scientific expeditions.

Having made these points, I have to insert a cautionary note. Sometimes the discussion of human precursors is cast in terms of a "missing link" — a creature midway between apes and modern humans. The notion of the missing link is based on the idea that we somehow descended from modern apes. This is a misreading of evolutionary theory, which teaches nothing of the sort. Instead, it teaches that both modern humans and the great apes descended from a common ancestor who lived millions of years ago.

Because of the problems outlined above, research on the earliest humans and the search for a common ancestor tends to move forward in sudden spurts, often separated by decades, as new finds are made.

It wasn't until 1994 that the discovery of a major pre-Lucy find, *Australopithecus ramidus* ("southern ape, root of humanity") in Ethiopia was announced. This animal is known to us through one forearm, part of a skull and jaw, and several teeth — parts of about seventeen individuals in all. Nevertheless, the evidence of these fossils (the canine teeth that are larger than those of other hominids but smaller than those of apes, for example) places *ramidus* squarely between Lucy and the earlier primates. This discovery (the first major find in the region since Lucy) reduced the gap of ignorance by a half-million years.

I expect that the uncovering of the human past will go on at this same slow pace, with new fossils showing up every decade or two. I don't expect any dramatic changes in the way fossils are collected or in the rate of discovery. The fact that *ramidus* didn't cause anyone to redraw the family tree, however, may mean that new finds in the future will be less revolutionary than those of the past.

How Fast Does Evolution Proceed?

WHEN CHARLES DARWIN first put forward his theory of evolution over a century ago, he thought that evolutionary change proceeded slowly and gradually, with small changes accumulating and eventually leading to the enormous diversity of species we see around us today. There was one problem with this view, a problem that Darwin himself recognized. If you look at the evidence contained in the fossil record, you often find situations in which one species lasts for a long time, then disappears suddenly, to be replaced by another species, and the pattern is repeated. The view that evolution takes place by the slow accumulation of changes has come to be called *gradualism;* the view that long periods of stability are interspersed with bursts of rapid change is called *punctuated equilibrium.*

You may be wondering why there is any debate about the process of evolution, since the evidence of the fossil record should tell which view is right. To see why sorting out the rate of evolution is such a problem, you have to understand a little about how fossils are found (or collected, to use the paleontologist's jargon).

If an organism is buried rapidly when it dies, it may go through a process in which water flowing through the soil gradually replaces the original bones and hard parts with minerals. The result: a reproduction in stone of the original part, which we call a fossil. If natural weathering brings the fossil to the surface and it is collected by a paleontologist, it becomes part of the fossil record.

As you might guess, this process is pretty erratic. An animal that dies on land, for example, is much more likely to have its bones scattered by predators than to be fossilized. In fact, scientists estimate that fewer than one species (*not* one individual) in 10,000 is even represented in the fossil record. Paleontologists find a few fossils of a species in one location and a few fossils somewhere else, then try to piece together a family tree. It is not at all uncommon for there to be gaps of hundreds of thousands or even millions of years between two appearances of a species in the fossil record. Obviously, in this sort of situation it would be easy to miss the change from one species to another.

Here's an analogy: if you looked at a picture of a busy city, you would find many people in cars and many people on foot. Chances are, however, that you would see in the photo very few people — perhaps none at all — actually getting into a car. Unless you had a number of detailed pictures of the city over a period of time, it would be easy to miss pedestrians becoming drivers. In the same way, unless you have an unusually full and detailed fossil record, it is easy to miss the transition from one species to the next. This, in fact, is the traditional Darwinian explanation of why the fossil record might look punctuated even though the underlying evolutionary process itself is gradual.

The only way to clear up the question is to go to places where the fossil record can be seen in unusual detail. One of the first such studies was done in some mountains in Wales where sediments that collected in a quiet bay hundreds of millions of years ago produced an exceedingly fine-grained set of rocks. In those rocks is the evolutionary record of trilobites. By carefully collecting and analyzing those fossils, British paleontologist Peter Sheldon was able to trace one species changing into another. His verdict: in this case the evolution was gradual.

But this isn't the last word on the subject. Scientists studying the evolution of coral-like animals called bryozoans over the past 15 million years have found exactly the opposite pattern. In their analysis, species of bryozoans stayed more or less constant for long periods of time, then produced new species in a matter of 100,000 years or so (a time, I should remind you, that is a mere blink of the eye by geological standards). This study, published in 1995, is the best evidence yet for punctuated equilibrium in the fossil record.

And so it goes. Paleontologists studying the evolution of prairie dog–like animals who lived in Wyoming 50 million years ago find gradual evolutionary patterns; scientists studying freshwater snails find some that evolved gradually over a 2-million-year period and others that seemed to exhibit a punctuated pattern. The correct answer to the question "Is evolution gradual or punctuated?" seems to be "Yes."

In the future this debate will be about two questions: (1) how often, and under what circumstances, do species exhibit gradual and punctuated evolutionary patterns? and (2) how do species remain stable over long periods of time in the face of changing environments? Both of these are interesting questions, and I, for one, am looking forward to a lively debate.

Will We Be Looking for Fossils on Mars?

IF MARS was really wetter when it was young — if there were really oceans on its surface, as we now believe — then a very interesting possibility arises. We now think that life on our own planet arose fairly rapidly after the period we call the "Great Bombardment," when large pieces of debris bombarded the newly formed planets. What if the oceans on Mars lasted long enough after this period for life to get started? Then there might have been life on our neighbor over 3 billion years ago, even if it was later wiped out. Could we find evidence for that brief experiment in life on the surface of the planet today?

Actually, life may have had more than one chance to start up on Mars. Not only was the planet wet at the very beginning of the solar system, when Mars was a young planet, but there is some evidence of intermittent periods of widespread flooding on the surface since then. In other words, there may have been periods (although none within the last billion years or so) in which bodies of water the size of the Mediterranean or even the Atlantic Ocean existed on Mars. It is possible that life developed several times on Mars, only to go extinct when the water disappeared. This means that although we cannot expect to find little green men (or even little green microbes) on the planet today, we might be able to find fossils if we looked.

On earth, scientists have found fossils of microbes — in fact, the evidence of single-celled bacteria constitutes some of the most important data supporting the theory of evolution. The basic technique involves finding old sedimentary rocks that were formed on ocean bottoms early in the life of the planet, cutting the rocks into thin sections, and examining the sections under a high-powered microscope. If you're lucky, you'll find impressions made by simple bacteria. On earth, evidence for 3.5 billion-year-old life has been found in this way.

On Mars, we can see areas of sedimentary deposits, which would be prime fossil-hunting terrain on the earth today. In addition, the fact that there is no plate-tectonic activity on Mars means that the rocks on the surface have been there a very long time; on the earth, in contrast, most rocks that were on the surface billions of years ago have disappeared beneath the ocean or have been weathered away. On Mars, with its thin atmosphere and lack of water, such weathering hasn't happened, and many old rocks are still available at the surface for study.

It's really very hard to overstate the importance of finding fossils on Mars. The questions that could be answered by finding a life form on another planet are so deep and so fundamental that they very seldom even get asked in modern biological sciences.

In essence, here's why it's so important. Every life form on the earth, from pond scum to the gray whale, is the result of a single experiment with life — a single genetic mechanism. Every living thing is descended from the same primordial cell, and this relationship shows up in our own cells. It's sobering to think that some stretches of human DNA have about a 50 percent overlap with the DNA of brewer's yeast — hardly a flattering point to make about *Homo sapiens*. Even if other forms of life started up on our planet at some time, no evidence of them seems to have survived. The entire science of biology, then, is based on the study of a single experiment — the one that took place on the earth billions of years ago.

But if life were to start again in different circumstances — if nature ran a different experiment — would that life be different or the same? Would it be based on carbon chemistry? On DNA? Would early fossils from Mars look at all like contemporaneous fossils on earth?

In the present context of biology, there's no way to answer these sorts of questions. If we found fossils on Mars, however, we would at least have the beginnings of an answer. If we found that the Martian experiment was based on DNA, as ours is, we could argue that there is some sort of inevitability about the molecular basis of life on earth. Such a finding would indicate that carbon-based life using DNA is, in some sense, the only form of life allowed by the laws of chemistry and physics. If, on the other hand, we found life based on some totally different chemistry, we would have to conclude that the mechanisms on earth are basically an accident — only one possibility among many. In this case, the entire science of biology would be seen as the study of an accident, and we wouldn't expect any life forms we encounter elsewhere in the universe to be like us at all.

But we won't be able to answer any of these questions until we find those Martian fossils. And here there's good news. The next Mars mission (set to arrive there in 1998) will land on a site where the first steps in the search — the study of minerals that might contain fossils — can begin.

What Ever Happened to the Neanderthals?

POOR NEANDERTHAL. From the very start he's been tagged as a loser, a shambling, stupid hulk who richly deserved extinction. Happily, these notions are changing, and modern science has been kinder to Neanderthal. We now recognize that the notion of a stooped posture and shambling gait came from analysis of the fossil of a man with advanced osteoarthritis! Neanderthal walked upright as we do — in fact, if you dressed him in a business suit and put him on the New York subway, he probably wouldn't provoke a second glance.

Neanderthal was actually our closest relative on the human family tree — indeed, some scientists claim that he was a subspecies of *Homo sapiens*. For reasons given below, I suspect that this isn't true, but the fact remains that he was a lot like us. Neanderthal buried his dead with ceremony, probably had a spoken language, and on average had a larger brain than we do.

There have been serious debates among paleontologists both over how the Neanderthals originated and why they disappeared. In 1992 an important discovery in a cave in Spain seemed to resolve the first question. The problem up to then had been that the few skulls found in Europe older than 130,000 years seemed to be very different from each other, and none seemed to have all of the standard Neanderthal characteristics (heavy brow ridges, elongated skull, long jaw). Some paleontologists argued that these remains were evidence for several different species of early humans, only one of which led to Neanderthal.

In the Spanish cave, paleontologists found a bonanza —

three well-preserved Neanderthal skulls dating to 300,000 years ago. These skulls, which obviously all came from the same population, were as different from each other as were all the previous skulls of different "species." The conclusion: wide variation in skull shape is a normal feature of the Neanderthal population, which can now be reliably dated back at least 300,000 years.

But the really interesting question about Neanderthal has to do with his relationship with early modern humans. Here recent studies of fossils found in caves in Israel are important. These caves appear to have been inhabited from 90,000 to 35,000 years ago, and beneath their floors are successive layers of fossils of Neanderthal interleaved with layers containing fossils of early modern humans. One way to interpret this data is to say that modern humans and Neanderthals lived side by side for 55,000 years without interbreeding — a situation that would argue strongly that Neanderthals were a different species from our early ancestors.

So what happened to them? The fossil record of Neanderthal is most complete in Europe, and there the story is clear. For several hundred thousand years the fossils of Neanderthal abound. Then, about 35,000 years ago, at a time corresponding almost exactly to the arrival of modern humans, Neanderthal disappeared from the fossil record. How this happened remains one of the great mysteries of human evolution.

In fact, scientific arguments on this question tend to be speculative in the extreme, for there is very little direct evidence — no Neanderthal skull with a modern human axe embedded in it, for example. It used to be assumed that Neanderthal was simply wiped out by the invaders — certainly a reasonable hypothesis to make in a century that has seen the likes of Adolf Hitler, Joseph Stalin, and Pol Pot.

More recently, however, as more data on Neanderthal

anatomy and lifestyles have accumulated, scientists have taken a somewhat different tack. Noting Neanderthal's large brains, scientists have started to ask why our ancestors were able to win that prehistoric competition. After all, if intelligence is an important factor in survival and if it's related to brain size, it should have been our ancestors that got wiped out, not Neanderthal.

Scientists now argue that although the brains of modern humans were slightly smaller than Neanderthal's, they were capable of different sorts of activity, particularly in the area of technological innovation. For example, for over a million years, the basic human tool was the hand axe — a stone laboriously chipped until it had a sharp edge on one side. Modern humans, on the other hand, perfected the art of stone flaking, a technique in which sharp flakes (each of which can serve as a tool) are knocked off a stone core. With this technique, modern humans were able to produce a wide variety of stone tools quickly. If Neanderthal really was unable to adapt to this kind of technological change, then there is no mystery about his disappearance. He simply represented one more species driven to extinction by someone better able to exploit a particular ecological niche.

Having said this, I have to add that in this area of science each new discovery (like the Spanish cave) can completely revolutionize our thinking. Don't be surprised, in other words, if the conventional wisdom changes drastically a few more times before settling down.

When Did Human Beings Arrive in North America?

AMERICA IS A continent of immigrants. All of us are descended from people who came to these shores fairly recently (geologically speaking). The ancestors of the Indian peoples of both North and South America came by foot from Siberia, and a major debate in archaeological circles these days concerns whether that migration took place 11,000 years ago or at a significantly earlier time.

We know enough about recent geological events to say with some precision when populations could and could not have entered North America. Some 25,000 years ago, glaciers started to grow near the North and South poles, locking up water in the form of ice and lowering the levels of the world's oceans. At that time it was possible to walk from Siberia into what is now North America across the Bering Strait. By 20,000 years ago, however, the glaciers had advanced far enough south that it was no longer possible to traverse the region by foot, and this situation lasted until about 12,000 years ago. At this point, there was a brief window of opportunity while the glaciers receded but sea levels remained low — a window during which the ancestors of American Indians could have come across.

The conventional wisdom among archaeologists for the last fifty years has been that the ancestors of American Indians came across during this second opening of the Bering Strait window. The reasoning is very simple: there are no authenticated and accepted archaeological sites in North America that indicate any human population was present before about 11,000 years ago.

In fact, the earliest known archaeological site on this continent — the site that gave its name to an entire culture — was discovered near the town of Clovis, New Mexico, in the 1930s. The "Clovis people" were characterized by the rather distinctive fluted shape of their arrowheads and other stone tools, and an archaeologist can tell you at a glance if a particular artifact belongs to them.

The conflict arises because of new findings by molecular biologists, who have analyzed the mitochondrial DNA of groups of living Native Americans (recall that the DNA of the cell structures called mitochondria is inherited only from the mother). The amount of difference in DNA sequences between any two groups is taken to be a measure of the length of time since those groups shared a common ancestor. In general, the data on mitochondrial DNA indicate that American Indians in the groups studied shared a common ancestor between 20,000 and 40,000 years ago. The molecular biologists, then, argue that the tribe carrying this DNA walked across the land bridge during the first window of opportunity and that they were the true ancestors of the American Indians.

This is where the conflict stands at the moment. One problem with the view that the migration happened 11,000 years ago is that Clovis sites have been found all over the Americas, and all are about 11,000 years old. For this to have happened, the ancestors of American Indians would have to have moved from Alaska to Tierra del Fuego (some 10,000 miles) in about three hundred years — about thirty miles a year. This is an extremely rapid rate of travel for a hunter-gatherer society that has to keep in touch with their kin, make sure everybody is fed, and maintain a community life. As one scientist joked, "Have you ever tried to go *anywhere* fast with kids?" Some scientists feel much more comfortable with the idea that people arrived 20,000 years ago and spread southward at a more leisurely pace.

On the other hand, if you take this point of view, you have to explain why, in spite of extensive searches, anthropologists have yet to find a site that everybody agrees is older than 11,000 years. There have been, to be sure, many candidates for such sites — typically, someone will announce a site as being pre-Clovis and then, on further examination, archaeologists will discover a mistake in the dating or decide that what was called a hand tool is a natural piece of rock. In the end the site is rejected. At the moment, there is only one site — a place called Monte Verde in southern Chile — that may be pre-Clovis. There is also a cave in Brazil that seems to have artifacts as old as the oldest Clovis site.

In the meantime, more studies of mitochondrial DNA are being done. These studies give a wide variety of dates for the last common ancestor, most likely because they analyze different parts of the mitochondrial DNA. Some of the analyses put the last common ancestor of some Eskimo tribes as far back as 80,000 years. In this situation, the ancestral tribe was almost certainly isolated in Siberia for a period of time before it came across the land bridge. In such cases, the ancestral "molecular clock" probably was started before the people crossed over into North America.

To tell the truth, I don't have any idea how this controversy will play itself out, but I'm reluctant to accept any arguments about early arrivals until someone can point to a place and say, "Look! Here's where people lived before Clovis."

Does Anyone Here Speak Nostratic?

 OUR LANGUAGES tell us a great deal about both the societies in which we live and our history. English, for example, includes words like *hand, shoe,* and *bread* (which are essentially the same as their German equivalents) as well as words like *pendant, fumigate,* and *cataclysm* (which are essentially the same as Latin and French); from this we can deduce a great deal about the history of England. In an analogous way, linguists in the distant future, seeing words like *microchip* and *airplane* in today's English, will be able to reach some accurate conclusions about our technical abilities.

Linguists have used sophisticated versions of such analyses to trace the evolutionary history of the world's languages. By studying existing languages and using well-known laws of linguistics, they deduce the earlier forms. One such law was formulated in 1822 by Jacob Grimm (better known for the anthology of fairy tales he wrote with his brother). The basic idea of Grimm's law is that there are patterns in the way sounds change from language to language — consonants like "b," "d," and "g" tend to be replaced by consonants like "p," "t," and "k," for example.

Using this sort of analysis, linguists have been able to reconstruct a good portion of Indo-European, the language that forms the base of most modern European and Indian languages. (The generally accepted notion is that Indo-European was spoken about 6,000 years ago.) The language contains many words for things like high mountains and mountain rivers, so the original people probably came from a mountainous region.

It also has words for grapes, barley, and cherries, which indicate both the practice of agriculture and (for the grapes, at least) a southerly location. The best bet for the location of the original language: eastern Turkey and the Caucasus Mountains.

This kind of reconstruction isn't just some ivory-tower dreaming, either. In the late nineteenth century, a French linguist using this sort of analysis argued that there had to be languages, now lost, with certain characteristics. In the 1920s, excavations in eastern Turkey turned up clay tablets written in a language he had predicted (which is a version of Hittite). These excavations, showing that several different Indo-European languages existed about 4,000 years ago, also yield information on when the original language started to split off into its many descendants.

So far, we are on ground that is widely accepted by linguistic historians. The conflict starts when people begin to try to work backward from Indo-European to a (possible) earlier language. Some Soviet linguists, in fact, claimed that they could trace back to a hypothetical language called Nostratic (from the Latin for "our"). Nostratic is supposed to be the base of Indo-European and also of Semitic languages like Arabic and Hebrew and Asian languages like Korean. In fact, proponents of the theory claim that about three quarters of the human race speak languages derived from Nostratic, which is supposed to have been spoken about 12,000 years ago.

Of course, this theory is not without its critics. The main opponents tend to be the scholars who worked out Indo-European. They do not claim that the notion of a language like Nostratic is wrong, but they are affronted that its proponents would put forward such a theory without going through the detailed and painstaking work that went into the Indo-European reconstruction. If the Nostratic theory is right, these critics will, presumably, be won over in time.

What does Nostratic tell us about the culture of those who spoke it? At the moment, linguists claim to have worked back to about 1,600 Nostratic root words. The language seems to have many words for plants but none for cultivated plants. The words for animals do not distinguish between domestic and wild varieties. There is even a word *(haya)* which seems to mean "running down game over a period of several days." Clearly, the speakers of Nostratic had not yet developed agriculture and lived as hunter-gatherers.

No one expects linguistic analysis to tell us the whole story of the early migrations of modern humans. Archaeological data, whether from written records or other artifacts, have to be found to back up any linguistic theory. The problem, of course, is that hunter-gatherers do not produce many artifacts, so their archaeological trail is very difficult to trace. Recently, molecular biologists have tried to collate studies of DNA with the linguistic theories to show that the story of the branching of peoples told by language is the same as the story written in DNA. This is a somewhat dubious undertaking, since people can change languages without changing genes and vice versa. Modern Hungarians, for example, are genetically European, but they speak a non-Indo-European language brought in by the Magyars over a thousand years ago.

Finally, a few linguists want to reconstruct parts of the original protolanguage that gave rise to Nostratic and *all* modern tongues. I think I'll wait a while on that claim.

8

Technology
(Mostly Computers)

If We Can Send a Man to the Moon, Why Can't We Make a Decent Electric Car?

 THE ANSWER, in a word, is batteries. A battery is a device that stores energy in a set of chemical compounds. When these compounds interact, they produce (among other things) a flow of electrons in an external wire, which we perceive as an electric current. In a lead-acid battery like the one in your car, for example, chemical reactions between lead and sulfuric acid at one plate free up electrons, which then run through an outside wire to hook up with atoms undergoing chemical reactions at the other plate.

When the original chemical compounds have done all their work, the battery is discharged and no more current flows. At this point, some types of batteries (like those in your Walkman) have to be thrown away. Others (like the one in your car), can be restored by running current (from a generator, for example) backward through the battery, recharging it. So the principle behind an electric car is simple: use batteries to run the car all day, then plug it in overnight and recharge.

The requirements that an electric car battery has to meet are pretty stringent, and it's not clear that we can meet them all. For one thing, the battery has to store enough energy for the car to have a reasonable range (the distance you can go without recharging). Most electric cars now have a range of sixty miles or so, but that will have to increase to over one hundred miles for the cars to be commercially marketable. The battery has to have enough power to accelerate the car quickly (the "golf cart syndrome" is a major bar to the sale of electric cars). It has to perform well for hundreds of cycles (ideally, you should never have to replace a battery pack during the car's lifetime), and it should recharge quickly. It has to operate in all temperatures and be essentially maintenance free. The battery should be cheap and, when its life is over, it should be recyclable. And that, my friends, is quite a list of requirements!

Because California has been discussing the requirement that "zero emission" (that is, electric) cars be sold in the state, there is a major drive to produce marketable electric cars right now, and a concomitant interest in battery technology. Planners differentiate between midterm goals (to meet the California requirements) and long-term goals (to replace the internal combustion engine). The midterm goal, roughly, is a car with a range of 100–125 miles and a cost for the batteries of about $5,000.

There seem to be three horses in the battery race. One is the old lead-acid battery, which is the cheapest, but it has problems meeting the energy and power-storage requirements. Nickel-cadmium batteries have been around for a long time, too, and are routinely used in industry and in consumer electronic applications; chances are the battery in your laptop is of this type. These batteries deliver about twice as much energy per pound as lead-acid and can go through many recyclings. Unfortunately, they cost about twice as much. Finally, there are batteries that operate on chemical reactions between nickel and a mixture of metal compounds. You may have seen some of these "nickel-metal-hydride" batteries in computer equipment recently. General Motors and a small high-tech company called Ovonics have been developing cars powered by these batteries, which store about two and a half times as much energy as a lead-acid battery and, as they go into production, have costs comparable to those of nickel-cadmium.

Ovonics has been running converted GM cars for several years in Michigan, so the company has data on real-life performance of their batteries. Their range is over 125 miles in the city (150 on the highway), and they seem to have plenty of zip. In an episode Ovonics doesn't like to publicize, one of their engineers took a car out on the freeway and was clocked at 80 miles per hour (though not, fortunately, by the police). It seems to me that if they can bring the cost down, this will be the battery that powers cars in the future.

A word about the market for electric cars: they will be used mainly for commuting and city driving — no one would want to take one across the desert. Peugeot has been testing a car with a nickel-cadmium battery in La Rochelle, France, for the past few years, and is so encouraged that it produced 5,000 electric cars in 1995. An interesting sidelight is that market studies find that people's main concern with electric cars is the fear of having the batteries run down while they're out on the road. In La Rochelle, people typically recharged their batteries every night, usually when they had about 70 percent of their energy left. The Impulse (a GM demonstration car with lead-acid batteries) has a "lock down" mode built in. When the batteries get low, the car first shuts off all auxiliaries (like air conditioners and radio) and then won't go over 20 mph. Engineers call this the "limp home" mode. My sense is that if many people are caught on the expressway in cars that won't go over 20 mph, it will be a while before the electric car becomes a hot item on the American automotive scene.

How Powerful Can Microchips Be?

IN THE 1960s, at the very beginning of the computer revolution, a man named Gordon Moore (who went on to become one of the founders of Intel) noticed that the growth of computer technology had an unusual feature: the memory of computers (the amount of information they can store) was doubling every two years. In the time since then, we have come to realize that almost every index of power in a computer — the size of circuits, the speed at which processing can be done, and so on — seems to get twice as good every two years as well. Collectively, these observations are often referred to as Moore's law.

Now, you should be very clear that Moore's law is not a law of nature in the sense that Newton's law of gravitation is. It is simply an observation that the technology associated with computing and information processing grows rapidly. What I find so striking about it is that this growth seems to be independent of the technology being used. For example, no one in the late 1960s could have foreseen modern microchip fabrication methods or methods of circuit design, yet the rate of growth at that time was much the same as the rate today, regardless of the intervening technological revolutions.

To give you a sense of this growth, we can talk about one measure of technological capability — the size of the smallest feature that we can put on a microchip. In the late 1960s, the smallest line we could make was about 20 microns across — roughly the distance across 200,000 atoms (a micron is 1 millionth of a meter). Since then, that size has continued to drop, reaching 5 microns (about 50,000 atoms) during the height of the integrated circuit boom in the late 1970s, 1 micron (about 10,000 atoms) during the personal computer/microchip era of the 1980s, and today stands at about .10 micron — the distance across several hundred atoms. I can't think of a better way of illustrating the fact that Moore's law is independent of technology — the *way* those lines are made has gone through several revolutions since the 1970s, yet the drop in its size seems steady and inexorable.

Given the kind of technological growth we have witnessed in the past, what can we say about the future of the computer? My own sense is that computers are unlikely to get a lot smaller than they are now, for several reasons. One is that the emphasis today is not on large mainframe computers but on small units that can be networked together. In this case, the limit to the size of the computer is set not by technology but by human physiology. For example, a laptop can't get much smaller than it is today because of the size of the human finger — at present the only way you can feed information into the machine is by typing. (Of course, this restriction might change if computers learn to follow voice commands, as they do in *Star Trek*.) In addition, the size of letters easily read by the human eye puts a limit on the size screen we can use to display the results of calculations. Consequently, in a practical sense, the limit to the size of the microchip will not be what defines the size and weight of laptop computers. The weight will be determined primarily (as it is now) by the weight of the battery, while the physical size will be determined by the size of the human finger and the acuity of the human eye.

In addition, where microchips are concerned we are almost at the point of running into fundamental physical limits. Take the example of the smallest feature we can inscribe on a microchip. Research programs are in place that will allow widths of a few hundred atoms, but at this level we start to encounter problems. It is very difficult, for example, to get electrons around sharp bends when the channel they're following is that narrow — they tend to leak out the side. These sorts of difficulties can be expected to multiply as the feature size is reduced. There is, in addition, an absolute limit at about one thousandth of a micron (about the size of a single atom), which Moore's law says we will reach in about 2060. It is impossible even in principle to have structures smaller than that on a silicon chip.

I find one feature of the growth of computer capability particularly fascinating. The size of the smallest feature you can make obviously affects many other quantities — the number of transistors that you can put on a single chip, for example. If you make reasonable assumptions about this relation and then extrapolate into the future, you find that sometime between 2020 and 2030, we will reach the point where we can put at least 100 billion transistors onto a structure the size of a bouillon cube. This is about the same as the number of neurons in the human brain. In other words, if computer technology progresses in the future as it has in the past, within the lifetime of many people reading this book, we will be able to produce (at least in principle) an artificial system of the same size (if not the same complexity) as the human brain.

What then?

Will We Be Computing with Light?

 ARE WE NOW reaching some sort of limit in the power of computers? This is such an important issue that I discuss different aspects of it in several other places. Here I want to talk about one particular strategy for increasing the power of computers and a particular technology that is almost ideally suited for it. The strategy is called parallel computing; the technology is the optical computer.

When an ordinary computer deals with a problem like analyzing a picture, it proceeds in a straightforward, logical way. It starts by analyzing a dot in the upper left-hand corner, say, then moves over and analyzes the second dot in the first row, the third dot, and so on until it works its way to the lower right-hand corner. In this system, called serial computing, the entire power of the machine is devoted to each piece of a problem, one step at a time. With a few important exceptions, all computers today operate this way, whether they are analyzing a picture or carrying out a complex mathematical calculation.

One reason that the human brain is so much better at analyzing images then the most powerful computers is that the human brain does not operate sequentially. In fact, what we know of human visual processing uses an entirely different strategy in which different parts of a picture are analyzed simultaneously by many small "computers," whose results are then passed up to the next layer of analysis. This is called parallel computing, and it is one important strategy for making machines faster.

Ordinary silicon computers can be made to operate in the

parallel mode — indeed, versions of such machines are already on the market. The main problem is that when you break a problem into little pieces and route those pieces through different parts of a machine, you have to spend a lot of time worrying about traffic control — making sure that part A finishes a calculation before part B needs the results, and so on. Some computer scientists expect that silicon machines using parallel computing will be able to perform a trillion operations a second (a sort of Holy Grail in the computer game) by the end of the century.

But in 1990, a new dark horse entered the parallel computing race. Called the optical computer, this machine substitutes beams of light for electrical current running through wires as its means of communication. The basic working element of the optical computer is a tiny glass tower containing layers of semiconductors. A laser beam shines on the tower and releases electrons that get trapped in the semiconductors. Depending on the strength of the laser signal, the electrons turn the tower either translucent or opaque. A stronger laser is then shone onto the system, and it will be reflected either strongly (from the translucent tower) or weakly (from the opaque tower). This reflected signal is one bit of digital information — yes or no, on or off. The reflecting tower plays the same role in the optical computer that the transistor does in the silicon machine (a transistor, recall, acts as a switch, which can either let current through or block it).

A prototype optical computer element would be an array of thousands of towers on a glass plate, together with lenses to focus the lasers — one set of lasers for each location on the plate. Thus the optical computer is inherently parallel, since there are always many operations going on at the same time. A full-scale computer would presumably operate by passing light through many such plates, then reading the light intensities of the laser beams that come out the other side, much as ordinary computers read electrical current. Advocates of this technology believe that they may have such a system working by the turn of the century.

There are a number of reasons to believe that optical computers will play a role in our future. To give just one example, as engineers make silicon microchips smaller and smaller, they run into a serious problem that is due to the nature of electrons. If electrical current is running through two wires that are very close together (as they must be in ultraminiature machines), then the electrons in one wire will start to affect those in the other. This limits the flow of information into the computer in much the same way that a tunnel or bridge limits the flow of traffic into a city. Beams of light, on the other hand, can travel side by side or even intersect each other without interference. So as our computers get better and better, they may also become very different from the ones we use today.

Must Computers Be Made of Silicon?

IN THIS DAY and age, everyone knows what a computer looks like, and many of us have a notion of how they work. The digital computer, after all, isn't all that different from an abacus or an adding machine, except that instead of shuffling beads or wheels, the computer turns transistors on and off by shuffling electrons. Even experimental technologies like the optical computers discussed earlier operate in essentially the same way, by turning switches on and off.

There are, however, some new schemes on the horizon that could produce "computers" that few of us would recognize — vats of chemicals, for example, or cubes of perfect crystals held at temperatures near absolute zero. In these devices, computing — the manipulation of information — is done entirely differently from the way the familiar digital computer operates. Whether they will ever be commercially important (or can even be made to "work" in any practical sense) is not clear at this time. If they do work, however, don't expect to see them for sale at your local "Computers R Us" store. Their use will almost certainly be confined to specialized applications on specific types of problems.

Calculating with DNA

The first "molecular computer" was put into operation in 1994 at the University of Southern California. It was used to solve a simple version of the classical computing problem called the "traveling salesman problem" or, more technically, the "directed Hamiltonian path problem." This problem can be stated as follows: given a set of vertices, one end of which is designated as "in" and one as "out," can you find a path that goes from in to out and visits every other vertex once and only once? As discussed elsewhere, this type of problem is called "NP complete" in complexity theory. These problems assume a special significance in mathematics, because if they can be solved, those solutions can be used to solve a host of other problems as well.

The molecular computer works this way: a number of single strands of DNA are prepared. Each strand is twenty bases long (recall that the ordinary double-stranded DNA molecule is like a ladder, with each rung made by joining together two molecules called bases). Each vertex is represented by one twenty-base molecule, and each possible path between vertices by another molecule in which ten bases are designed to hook on to half of one vertex, and ten are designed to hook on to half of a different vertex. Billions of these molecules are dumped into a container and allowed to combine with each other. A "path" molecule will hook on to two vertices, leaving half of each vertex molecule to hook on to another path molecule. Once a path attaches to a particular half of a vertex, no other path can bind at that point, guaranteeing that each vertex has only two paths leading to it. In the end, chemists search for a molecule in which the "in" vertex and "out" vertex are free and all other vertices appear only once. This molecule codes the solution to the problem.

The DNA computer works because the molecules combine in all possible ways, most of which do not represent solutions to the problem. Only combinations that "work" (if any) are picked out at the end. In a sense, the molecules are carrying out a massive exercise in parallel computing, and the solution to their calculation is contained in the way the successful molecules hook up.

Quantum Computers

The basic workings of the quantum computer depend on one of the strange properties of quantum mechanics — the notion that a given particle can be thought of as existing in many different states at the same time and that only the act of measuring it "forces" it into one specific state. Think of each particle as being analogous to a wave on water, where the height of the wave at any given point is proportional to the probability that the particle will actually be at that point — if the wave peaks at a particular point, it means the particle is likely to be found there.

A quantum computer would be set up so that all possible waves move through the system, but only those waves representing the right answer will reinforce each other at the end, while those representing the wrong answers will cancel each other out. Such a computer might be a specifically designed crystal in which the waves are represented by different vibrations moving from atom to atom. The "problem" might be fed into this computer by having laser beams trigger vibrations on one face of the crystal, and the answer could be read by monitoring vibrations on another face.

This theoretical scheme has attracted its share of critics. The main practical problem seems to be that quantum waves can be distorted by imperfections in the crystal, just as waves on water can be distorted by a rock sticking up on the surface. Thus even a microscopic defect could mess up the quantum computer. On the other hand, in 1996 scientists managed to operate one piece of a quantum computer — a so-called logic gate. I guess we'll just have to wait on this one.

Can We Make Computers
That Learn?

JUST BECAUSE the brain can compute doesn't mean it's a computer. Despite some superficial similarities, there are enormous differences between the "wetware" inside your skull and the hardware inside your computer. The basic unit of the computer is the transistor, which typically receives an electrical signal and, on the basis of that signal, goes into either an "on" or an "off" state. The basic unit of the brain is the neuron, which takes in signals from thousands of other neurons and, by a complex process we don't yet understand, integrates those signals and either "fires" (sends nerve signals out to many other neurons) or doesn't.

Another difference is that once a conventional computer is wired and given its instructions, it continues to do exactly what it is told — indeed, the ability to repeat complicated tasks over and over is one of the great strengths of the machine. A brain, on the other hand, changes and evolves as it is used, learning new tasks in the process. It does this, in part, by changing the strength of the connections between neurons.

Computer scientists have made important advances by designing machines that function more like the brain. The most important development involving this type of research is the "neural net." This kind of network has three parts: an input system that takes in information, a processing system that deals with the information and that can be modified, and an output system that reports the results of calculations. By a process I'll describe in a moment, a neural net can actually learn to carry out operations that are not programmed into it at the beginning.

Let's take a simple example of a learning task — recognition of a pattern. The input section of the neural net might be a set of phototubes focused on the screen of a black and white TV set. In this case the picture will be broken up into tiny squares (called pixels), one pixel per phototube, and the light coming from each square is measured. The input signal to the system would then be a series of electrical currents, the strength of each corresponding to the lightness or darkness of a specific pixel. (For reference, the picture in a standard American TV set is split up into pixels by breaking each side into 525 divisions. Thus, there are $525 \times 525 = 275,625$ pixels in the picture.) We can compare the input section of a neural net in human beings to the cells in the retina that first convert light to nerve impulses.

The signals then go to the processing unit, and it is at this point that a neural net starts to differ from an ordinary computer. Suppose that the task is to read a number on the screen. As information from the input starts to work its way through the processing system, a series of rules will be followed about how much weight to give each signal. For example, pixels at the edge of the picture might be less important than those at the center. The processing units, then, might be told to weight pixels from the center twice as heavily as those from the edge. Farther along there might be similar rules about edges ("weigh the signals where everything to the right is dark and everything to the left is light three times as heavily as those where light and dark are intermixed") and other features that are important in recognizing numbers. In the end, a signal will go to the output, where one of ten units will fire, corresponding to the number that is read.

At this point, the learning starts. Suppose you feed in the

number 2, but the net tells you it is 7. You tell the computer that it's wrong, and it is programmed to go back and rearrange the weightings in the processing unit — it might, for example, decide to weigh pixels at the center four times as heavily as those from the edge instead of twice. The system would run again, check the result, readjust the weight, run again, and so on. In the end, the net will have learned to read the numbers off the screen.

Reading a single number is a relatively simple task, but systems like this can do much more complex tasks. For example, they can be trained to recognize acceptable patterns in materials moving down a production line and thus play a role in industrial quality control. They can be trained to recognize individual faces and voices in security systems. There are even systems in the final stages of testing that can be trained to read handwritten zip codes on letters and thus speed mail sorting (for handwriting like mine, this is no small task).

At a higher level, sophisticated versions of these neural nets can analyze large amounts of data and extract patterns that are much too complex to be seen by human observers. Such machines can be very useful in understanding systems that have a large number of interdependent variables, such as agricultural ecosystems or large molecules.

Having said this, however, I have to make an important distinction between two questions: whether neural nets are useful (they surely are) and whether they provide a model for the workings of the human brain (which remains very much an open question). My guess is that the answer to the second one will turn out to be no.

How Small Can an
Electrical Circuit Be?

 THE FIRST transistor was about the size of a golf
ball. Today, choosing from a variety of standard-
ized techniques, engineers can easily put millions
of transistors on a single microchip no bigger
than a postage stamp. By the beginning of the next century,
that number will surely be higher.

Every time the number rises, the size of electronic devices
shrinks and we take another big step in the information revolution. When I was a graduate student in the late 1960s, I did
calculations on what was then a state-of-the-art computer at
Stanford University, in the middle of the nascent Silicon Valley. The machine took up a whole room and required a crew of
several people. You fed in instructions on a stack of cards and
had to wait several hours to get results. Yet that huge machine
had considerably less computing power than a modern laptop,
which could handle my thesis calculations in milliseconds and
write the thesis in its spare time. Making circuits smaller, then,
can produce revolutionary effects in our lives.

But here you run into technical problems. In order to put
more circuits on a chip, scientists will have to start making
transistors and circuit elements that are comparable in size to
a few atoms (today's circuit elements are several thousand
atoms across).

The standard miniaturization techniques in use today are
simple to describe. In one common process, a silicon substrate
is coated with a varnish-like material called a "resist," and a
pattern is etched into it. For example, a mask may be placed
over selected areas before light is shone on a resist made of a

particular chemical that undergoes changes because of the light (much as photographic film does). Other chemicals may then be used to remove the unchanged parts, leaving behind extremely fine lines and patterns in the resist.

Obviously, making miniature circuits requires the ability to draw very fine lines. This is normally done today by using a "pencil" made of high-energy electron beams. Such beams can produce lines in the silicon that are some twenty to forty atoms wide. You may have seen photographs of company or agency names written in lines a few atoms wide. This seems to be a favorite hobby of the guys and gals in the lab — it obviously gets their administrator's attention.

There are, however, limits to what you can do with high-energy electron beams, and scientists are hard at work developing ways of drawing even finer lines. Let me tell you about just one experimental technique, involving a completely new kind of "pencil." In this technique, scientists shine lasers across the surface of the substance they want to etch, then direct a beam of atoms toward it. The electrical forces of the light push the atoms into the troughs of the laser waves, leaving the crests unpopulated. This is similar to what you might see on a river, where flotsam collects in stagnant regions while the fast-flowing water remains clear. When the atoms come through the laser light they are in extremely fine lines at positions corresponding to the troughs of the laser beam. In this way, scientists at IBM and the National Institutes of Standards and Technology have succeeded in drawing lines about ten atoms thick on various kinds of materials. I expect that by the time you read this, they'll have made lines only one atom thick.

The advantage of this technique from a commercial point of view is that it can be used on any kind of solid material. For example, I saw a slide of grooves that had been cut into the surface of chromium. These grooves were about twenty atoms apart, a few atoms wide, and about three or four atoms high. They looked regular and clean, like the lines you make by running a fork through the icing on a cake.

Having noted this new technical virtuosity, however, I have to warn you not to expect immediate changes in your home electronic systems because of improved miniaturization. A whole constellation of technical problems has to be overcome before circuit sizes can be significantly reduced, and it's these problems that you will be reading about in the years to come. For example, when circuit "wires" are down to atomic dimensions, the electrons begin to behave strangely. Even now, with wires "only" a hundred atoms across, designers have to be careful not to have sharp turns in circuits, which let the electrons rattle around and leak out.

Furthermore, there is a strong sentiment in the technological community that the next frontier in computer building does not lie in miniaturization at all but in finding new ways to put circuit elements together — what is called computer architecture. Thus the ability to make very fine lines on our microchips does not necessarily translate into a technological breakthrough like those that produced the personal computer and the present generation of electronics.

Where Will Nanotechnology Take Us?

 IN THE KIND of machine shop that's been around since the Industrial Revolution, if you decide to make a gear, say, the procedure is simple: you take a block of metal and remove all of the atoms you don't want, leaving behind those atoms that make up the gear (to be honest, I doubt that many machinists think of their work this way). Scientists are just starting to develop a new way of manufacturing. The field of nanotechnology is devoted to understanding how to manipulate materials at the level of individual atoms and molecules. Instead of starting with a lot of atoms and removing the ones you don't want, the so-called nanotechnologists will build their structures by adding only those atoms that are desired, one at a time. A word about definitions: "nano" is the prefix for one billionth.

A nanometer is one billionth of a meter — the size of ten or so atoms lined up next to each other, or about one hundred-thousandth the width of a human hair. The term "nanotechnology" is often applied to any technique that produces very small things, such as the miniaturized circuits discussed elsewhere. Using techniques like those used in making circuits, scientists have succeeded in building incredibly small equipment — a completely functioning steam engine only a thousandth of an inch on a side, for example. But these techniques usually involve removing unwanted atoms (albeit on a very small scale) rather than adding them.

One of the most intriguing examples of nanotechnology involves powdery, crystalline, natural materials called zeolites.

The interiors of these crystals are full of nanometer-sized holes, or pores, of remarkably uniform width connecting larger chambers, or voids. The idea is that these pores will allow through only molecules or atoms of a specific size — in fact, the first industrial use of zeolites was as a kind of "sieve" to sort out different molecules. By controlling the atoms or molecules that are allowed to interact in the crystal, chemists use it as a set of "nano–test tubes" in which they can control reactions with unprecedented precision.

Using artificially produced zeolites, scientists at Purdue University recently succeeded in building "molecular wires" — electrical conductors only a nanometer across. Such wires could be used as connectors on highly miniaturized microchips. Another use for these materials is in chemical detection. A thin film of zeolites that will pick up only one type of molecule is laid on another material, and if that molecule is present in the environment, it will enter the tubes in the zeolite, changing the properties of the zeolite film slightly. In this way, even tiny amounts of the target molecule can be detected.

Another important area of nano-research is "designer solids" — large-scale materials made from specific molecular modules. Here the challenge is not so much in assembling the modules but in designing them so that they will assemble themselves into larger patterns. It would be like engineering bricks in such a way that when you threw them into a pile they would latch on to each other and produce a house. This task isn't easy, but it isn't as impossible as it sounds. Molecules and groups of molecules exert electrical forces on each other, pushing and pulling their neighbors around. If you can design the modules so that when a few are stuck together they pull in more of their kind, self-constructing materials might indeed become a reality. In the words of one researcher, "The whole point is to make a structure where the atoms are happy." For molecules, happiness equals growth.

A place where nanotechnology seems to be progressing rapidly is in the construction of so-called carbon nanotubes. When placed in a strong electric field, carbon atoms can arrange themselves into sheets, which then fold up into a set of nested tubes. A set of tubes typically will contain up to twenty individual tubes that are up to 20 nanometers across and some thousands of nanometers long. Some researchers are trying to find ways of keeping carbon atoms happy enough to grow nanotubes several feet long. If they succeed, they will have made the strongest materials known, either in nature or in the laboratory.

One application that may be commercially produced soon is nanometer-sized whiskers for use in lightweight composite materials. These materials, used in everything from aircraft to automobiles, consist of thin whiskers of strong material embedded in a ceramic or plastic matrix. They are the strongest lightweight materials we know how to make.

Nanowhiskers are made by exposing carbon nanotubes to a gas containing silicon, titanium, or some other metal. The tubes trap this gas, which then reacts with the carbon to make tube-sized whiskers of silicon or titanium carbide, the materials normally used in composite materials. Nanowhiskers, however, are only one thousandth the size of ordinary whiskers, which means that for a given weight of whiskers there is a lot more surface area to bind them to the matrix. The hope is that materials made from nanowhiskers will be even lighter and stronger than those currently in use.

How Virtual Is Your Reality?

SOMETIMES I THINK that the main characteristic of the information age is oversell. Certainly, after going through all of the breathless prose about artificial intelligence, for one, we should be a little skeptical of extravagant claims for the emerging technology of virtual reality.

The idea behind it is simple and comes from a basic principle of information theory. We know about the world through sense impressions, and every sense impression can, in principle, be broken down into an electronic signal sending units called bits (a contraction of "binary digit"). A bit is the answer to one simple question — on or off? up or down? 0 or 1?

Take your TV set as an example. Those pictures are produced by beams of electrons hitting a light-emitting coating on the back of the screen. The screen is split up into little squares called pixels (for "picture elements") — in American television, there are 525 divisions on each side of the screen, or $525 \times 525 = 275{,}625$ squares in all. Each pixel emits a controlled amount of light, and your eye integrates the resulting collection of dots into a smooth picture. You need 10 bits in order to differentiate between 1,000 different levels of light intensity in each pixel. A single black and white TV picture, then, can be specified by $275{,}625 \times 10$ bits of information. Want color? Repeat the calculation for each of the three primary colors, and your eye will integrate everything into a continuously colored picture. A colored picture, then, can be produced by a string of about 8 million bits of information — about the same number of bits it would take to code all the letters of this book in a digital computer.

A similar process of analysis can be carried out for sound (the bits represent the pressure of sound waves in successive instants of time) and touch (the bits represent pressure on the skin's receptors at neighboring points). In fact, *any* sensory experience, from a symphony to a large screen movie, can be represented by a string of bits. The only limit on what you can do is the rate at which you can process bits of information. (I'm not kidding about this. A guy at the University of Washington recently calculated the information content of "a completely satisfactory sexual experience" — turns out it requires transmission rates about a billion times higher than our systems are capable of now.)

So let me ask you this: what difference would it make if, instead of sitting and reading this book, you were wearing gloves that exerted just enough pressure on your hands to reproduce the feel of the book and were wearing goggles that took up your whole field of vision and reproduced what you're seeing now. If you realize that with enough information flow it would make absolutely no difference, you have grasped the promise of virtual reality.

If all that's needed to produce an experience is a flow of information, then we can use virtual reality to take us where we can never go in person — into an atom, for example, or the interior of a star. A familiar example of this idea is the modern flight simulator, in which a pilot can confront unusual situations and "crash" the airplane a dozen times while learning to get it right. An intriguing early use of virtual reality allowed chemists to grab virtual molecules and try to force them together. You could actually feel the forces the molecules exert on each other — when you had them lined up right, they would pull together like magnets.

Simple virtual reality systems allow scientists to sit in their laboratories and control robots descending to the ocean floor or into a volcano. The commercial market for virtual realities begins with games, but you can imagine all sorts of other uses — virtual business meetings indistinguishable from the real thing, travel to exotic places (or even to work) without leaving your home, and so on. The possibilities are endless.

But if that's true, why did I start these remarks on a cautionary note? The reason is that the reality systems we are likely to see in the near future are distinctly underwhelming. I have never seen a virtual reality display that a child couldn't tell from the real thing. Let's face it: our brains are pretty good at dealing with visual fields, and we're going to have to work pretty hard to fool them. Take three-dimensional displays as an example. Inside a computer, they are represented by lots of little polygons — some 80 million are required for a realistic image. You require a minimum of ten changes per second to perceive realistic motion (TV normally uses thirty). This means that to produce a believable three-dimensional image of "reality," a computer has got to process 800 million polygons per second. The best rate that's likely to be commercially available in the next decade is 2 million per second — less than 1 percent of what's needed.

This kind of disconnect between virtual reality and real reality abounds, and it's one reason I'm not going to expect big things from virtual reality anytime soon.

Why Do We Face a
Continuing Software Crisis?

 YOU MAY REMEMBER that when the new Denver airport finally opened in February of 1995, it was a year and a half late and the better part of a billion dollars in the hole. Why? Because the automated baggage system just didn't work. Why not? Because the instructions being given to the computers that ran the system couldn't handle the job of moving all those suitcases from one place to another. This was just one recent (and costly) example of a continuing problem we face as we incorporate computers into our everyday life.

Computer people distinguish between hardware (the actual physical machine, including the circuits on the microchip) and software (the instructions given to the machine when it has to carry out a specific task). For example, I am writing these words on a desktop computer — the hardware — onto which I have loaded a word-processing program — the software. The more complex the task to be performed, the longer and more convoluted the set of instructions has to be, and the more opportunities there are for mistakes to creep in.

And creep in they do. Denver airport is only a highly visible example of a system plagued with software problems. In 1993 California pulled the plug on a project designed to merge the state's files on driver's licenses and motor vehicle ownership. Cost: $44 million and seven years. In 1992 American Airlines, which pioneered modern computerized airline reservation systems, gave up on an attempt to merge their system with car rental and hotel systems. Cost: $165 million. And even when work is carried to completion, bugs hidden in the programs can surface in the most unexpected ways. The software on the space probe *Clementine,* after guiding a flawless survey of the moon's surface, suddenly starting firing the craft's steering jets, depleting the remaining fuel supply in a matter of minutes. As a result, the second half of the mission (a visit to a nearby asteroid) had to be scrapped.

Why should writing computer software, something we've been doing for half a century, still be such a headache? For one thing, software is still written by individual artisans, and every new set of instructions is, essentially, started from scratch. Like handcrafted furniture, each program is developed for a specific job. This means that the same mistakes can crop up over and over again, because correcting them in one system doesn't mean they'll be corrected in another.

Another reason is that the nature of computing is changing. Instead of creating programs for a single giant mainframe computer, software writers are working with systems that are widely distributed. The Denver airport baggage system, for example, had 100 computers dealing with input from 5,000 electric eyes, 400 radio receivers, and 56 bar code scanners. Techniques appropriate to one computing style may not be appropriate for the other, just as a carpenter skilled at erecting single-family homes may not be the best person to call on to design a skyscraper.

And, finally, computer programs are getting more complicated and ubiquitous. Sets of instructions many millions of lines long are not at all uncommon. Even a modern car's onboard computers may contain 30,000 lines — a length that would have been considered prodigious a few decades ago. Furthermore, the incursion of "smart" systems into ordinary consumer electronics means it is even more important to get the program right the first time — would you believe that even an electric shaver may have a chip running hundreds of lines of instructions?

Over the next few years, I expect to see American industry making a major effort to tackle the problem of producing reliable software to run our computers. Some lines of attack:

- The development of "mass production" techniques — programmers will develop modules of well-tested code that can be used over and over again in different programs.

- programs will be "grown" — small modules will be developed and tested, then put together into a final configuration, rather than trying to write everything at once. This process may also include "autocoding" — programming computers to write code themselves.

- Universities will offer much more extensive degree programs in software engineering, so that it will be possible to talk about national standards for the profession similar to those that exist in other kinds of engineering.

- Eventually, mathematical models will be developed to allow engineers to estimate the reliability of large computer codes, something we can't do very well now.

But no matter what we do, we will never make our computers or their software error-free. In the end, as with all other human endeavors, we'll just do the best we can and patch up the mistakes as we find them.

Where Next with Fuzzy Logic?

 IN THE WORLD of mathematics and logic, we are used to statements that are either true or false. Two plus two equals four — not usually four or about four but four, period. It is this kind of hard-edged logical statement that we normally feed into our computers and that computers normally manipulate.

Yet in our everyday world, we often make statements that aren't absolutely true or absolutely false but somewhere in between. For example, if I say that Dianne was sort of early for the meeting, no one would have any trouble interpreting the remark. Dianne showed up early, but not too early. She wasn't there an hour ahead of time, nor did she walk in the door as the meeting started. We can make sense of this statement without having to specify the time of Dianne's arrival exactly.

Welcome to the world of fuzzy logic. Conventional logic obeys something called the rule of the "excluded middle," according to which everything is either A or not A, white or black. Fuzzy logic deals with the rest of the world — a world in which shades of gray predominate. Since the 1950s, mathematicians have been hard at work on the rules that govern fuzzy statements, and their logic is now as well thought out (although different from) the logic to which we are accustomed.

If all I had to say was that there is another sort of logic in the world, it would hardly be worth discussing the subject in a book like this. However, fuzzy logic has enormous implications in all sorts of technologies. The way we control machinery, from factories to cameras, depends on rules that are much more easily accommodated in the framework of fuzzy logic than the more conventional sort.

When you are driving down the street, for example, you are constantly feeding more or less gas into the motor to maintain an appropriate speed. If this operation were to be turned over to a computer running a fuzzy logic system, we would start by making a set of rules (about which more later). The rules might say "If you're going too slow, put the gas pedal to the floor" and "If you're going too fast, take your foot off the gas." These rules would then be translated into fuzzy logic statements. On a city boulevard, for example, we might say that a speed of 25 miles per hour was neither fast nor slow but rather 70 percent slow and 30 percent fast. This weighting would then be used to calculate precise instructions for the engine — the computer might signal that the engine speed should increase by 286 rpm, for example. In addition, we could make the system more sophisticated by taking other factors into account. Traffic density, road conditions, and weather, for example, could all be dealt with in a similar way and weighted in the final decision.

Fuzzy logic systems have the enormous advantage of being relatively cheap. They provide solutions that are "good enough" without having to go through the enormous effort (and expense) involved in working out exact solutions. This approach has enjoyed great commercial success since the late 1980s, and it is estimated that it will be an $8 billion business worldwide by 1998. The subway system in Sendai, Japan, for example, uses fuzzy logic in its automated cars, which seem to provide a smoother, quicker, and more energy-efficient ride than cars operated by humans. Fuzzy-logic washing machines monitor the amount of murkiness in the wash water and adjust the length of the wash cycle accordingly. Fuzzy logic controls fuel flow and transmissions in some new model cars as well.

But you are most likely to encounter fuzzy logic in the operation of modern "point-and-shoot" cameras, in which sensors monitor the clarity of several areas of the image. Instead of demanding that the picture be perfectly in focus everywhere, the logic system sets the lens so that the focus is good enough in the test areas. These systems work so efficiently that fuzzy-logic camcorders can tell when the operator's hand is moving (as opposed to the object being filmed) and compensate for the movement.

The key point in building a fuzzy logic system is working out the rules to translate sensor readings into instructions. At the moment, these so-called expert rules are developed by a rather complex kind of engineering. In the future, however, they may well be developed by computers, using the neural nets discussed elsewhere in the book. A set of rules and some data will be fed into a computer, which will adjust the rules according to how well the output matches expectations. The new rules will then be tried, adjusted, tried again, and so on until the optimum set is obtained. This development will make fuzzy logic systems even cheaper and more efficient than they are now.

But having said all this, I have to confess that I think fuzzy logic represents what I call "transmission technology." Like the transmission of your car, it is a marvel of engineering and absolutely essential to the running of a modern industrial society. It is not, however, something that most of us have any deep desire to know much about, or something that would change our view of the universe if we understood it.

Will the World (or the Federal Government) End on January 1, 2000?

 IN A BOOK filled with discussion of serious issues such as the ozone hole, greenhouse warming, and the new virulence of viruses and bacteria, I thought it might be pleasant to end with a "doomsday" scenario that is perhaps not quite so drastic in its implications. I am referring to the coming end of the millennium and the effect it is going to have (and, indeed, has already had) on the operation of computer systems run by the federal government. You can regard this subject, if you like, as one more example of an unintended and unanticipated consequence of an obscure technological decision.

In this case, the decision was made decades ago and had to do with computer memory. In the days when new computer systems were being installed all over Washington, D.C., computer memory was a precious thing. Programmers and systems designers would go to extraordinary lengths to avoid clogging the memories of machines they were installing, because by today's standards, the memory capacities of those machines were pitiful.

Having been involved with computers throughout most of my professional career, I have vivid recollections of the memory problems in those days. It was not at all unusual for the best computers in major scientific laboratories to be incapable of dealing with the entirety of what today would be a rather mundane and trivial calculation. Often you had to start the numbers running, produce some intermediate results, then store those results on rolls of perforated paper. You would then empty the machine's memory, install the second half of the program and, finally, feed your first results in so that the calculation could be finished. For today's kids, raised on machines capable of storing huge amounts of information, this sort of computing is the technological equivalent of a Stone Age hand axe.

In any case, one of the memory-saving schemes that technicians installing federal computers in the 1960s hit on was to use only two-digit dates. For example, rather than using four digits to indicate the year 1965, they would simply write a 6 and a 5. This simple little trick saved enormous amounts of space in large databases. Think, for example, of a bureau of vital statistics not having to record the 19 in front of everybody's date of birth.

Obviously, this scheme would cause problems at the end of the twentieth century, when the first two digits of the date would change. Technicians in the 1960s, however, reasoned that by the 1990s we would be using new technologies, completely different programs, and totally revamped databases. In any case, they argued, thirty years in the future no one would still be using the same databases and the same programs they were installing.

Well, times have changed. The hippie movement has come and gone, disco has come and gone, and yuppies are now an endangered species, but those same old databases and old programs still constitute the heart of the federal government's computing system. The reason for this is very simple. It is always easier to modify an existing program than to write a new program from scratch. Each time the programs and databases for the federal government were updated and new systems were added, the old ones were not changed. In order to make new data consistent with what was already there, the two-digit date was adopted throughout the system.

The first intimations of disaster for most federal agencies

occurred in 1995, when people began asking for five-year projections on data by typing in a date like 3/25/00. If they were lucky, they got an error message on their screens saying, in effect, "I haven't the vaguest idea what you're talking about." If they weren't lucky (and there are some real horror stories making the rounds these days) the computer interpreted the last two digits of the year 2000 as a command to pull out data from 1900. I wonder how many laws and regulations are going into the books based on calculations in which this mistake was made and not caught.

To prepare for the next century, then, the federal government is going to have to hire a huge number of technicians to go through every database and every piece of software and change the amount of space allotted for dates from two to four digits. This is not a trivial undertaking. The Social Security Administration, which for obvious reasons was one of the first to run into this problem, required seven years to change its database alone — never mind the software. Estimates of the cost to the entire federal government for making this type of conversion: a cool $75 billion.

In fact, in this entire picture there is only one ray of consolation. January 1, 2000, will fall on a Saturday!

Table of Connections

The Physical Sciences

Astronomy and Cosmology

Biology (Mostly Molecular)

Evolution (Mostly Human)

Index

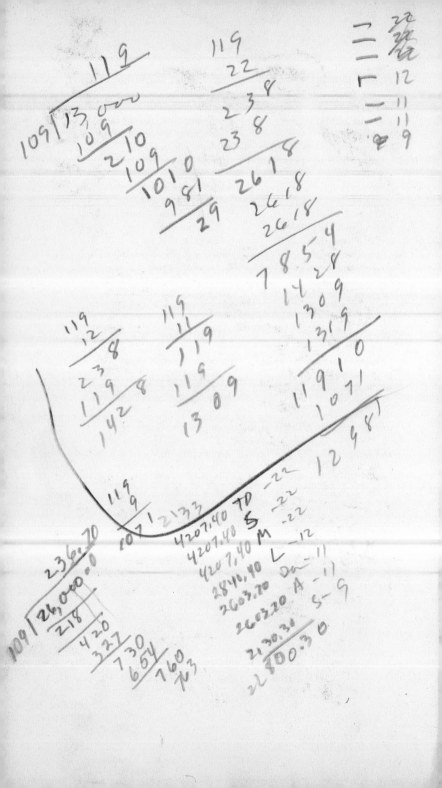